Refrigerant Management

Refrigerant Management
The Recovery, Recycling, and Reclaiming of CFCs

Billy C. Langley

Delmar Publishers Inc.™

I(T)P™

NOTICE TO THE READER

Publisher does not warrant or guarantee any of the products described herein or perform any independent analysis in connection with any of the product information contained herein. Publisher does not assume, and expressly disclaims, any obligation to obtain and include information other than that provided to it by the manufacturer.

The reader is expressly warned to consider and adopt all safety precautions that might be indicated by the activities described herein and to avoid all potential hazards. By following the instructions contained herein, the reader willingly assumes all risks in connection with such instructions.

The publisher makes no representations or warranties of any kind, including but not limited to, the warranties of fitness for particular purpose or merchantability, nor are any such representations implied with respect to the material set forth herein, and the publisher takes no responsibility with respect to such material. The publisher shall not be liable for any special, consequential or exemplary damages resulting, in whole or in part, from the readers' use of, or reliance upon, this material.

Cover Photo: G. Motil/Westlight

Delmar Staff
Senior Administrative Editor: Vern Anthony
Project Editor: Eleanor Isenhart
Art Coordinator: Cheri Plasse
Production Coordinator: Karen Smith

For information, address Delmar Publishers Inc.
3 Columbia Circle
Box 15015
Albany, NY 12212-5015

Printed in the United States of America
Published simultaneously in Canada
by Nelson Canada,
a division of the Thomson Corporation

10 9 8 7 6 5 4 3 2 1 xxx 00 99 98 97 96 95 94

Library of Congress Cataloging-in-Publication Data

Langley, Billy C., 1931–
    Refrigerant management : the recovery, recycling, and reclaiming of CFCs / Billy C. Langley.
      p. cm.
    Includes index.
    ISBN 0-8273-5590-4
      1. Chlorofluorocarbons—Environmental aspects   2. Chlorofluorocarbons—Safety measures.
3. Refrigerants—Environmental aspects.   4. Refrigerants—Safety measures.   5. Air conditioning.   6. Refrigeration
and refrigerating machinery.
      I. Title.
    TD887.C47L36   1994
    621.5'64—dc20   92-40577
                                                                                       CIP

CONTENTS

vii

Preface

Refrigerant Management: The Recovery, Recycling and Reclaiming of CFCs is designed to provide the latest information possible on the Clean Air Act of 1990, including the information necessary to meet the requirements set forth to this date.

The performance based objectives at the front of each chapter along with the summary, key terms, and review questions at the end of each chapter provide the reader with an indication of the minimum amount of knowledge that should be gained from each chapter.

Each chapter deals with a specific area of the refrigeration and air conditioning industry as determined by EPA for licensing requirements for refrigerant handling. Part I consists of six chapters that present the ozone depletion theory, the causes of the depletion, the theoretical solutions to the problem, and the basics of the Clean Air Act of 1990. Part II consists of two chapters dealing with the areas of refrigerant handling licensing. Chapter 7 presents low-pressure systems, high-pressure systems containing more than 50 pounds of refrigerant, and some of the high-pressure systems containing less than 50 pounds of refrigerant. Chapter 8 presents the remainder of the high-pressure systems containing less than 50 pounds of refrigerant and small appliances. The major reason for dividing the section on high-pressure systems containing less than 50 pounds of refrigerant is because of size and application of these units. Part III consists of one chapter that presents recovery/recycling equipment types, operation, and maintenance.

Because of the different areas presented for licensing purposes, there is some duplication of material. This was planned so that the reader would not always be flipping through pages to find the desired information and losing concentration. The reader can study the part that applies to a particular area of interest or all of the material for multiple licensing.

Upon completion of the study of the desired areas and the areas required by EPA for licensing, the reader should have the knowledge and confidence necessary to pass the exams and properly, efficiently, and responsibly handle refrigerants as set forth in the Clean Air Act of 1990.

Go forward with best wishes for a happy, healthy, and prosperous future.

Billy C. Langley

Acknowledgments

I would like to thank the following companies, agencies, and persons for their help and contribution to the making of this book. Without their help this project would have been much more difficult, if not impossible, to complete.

ARI (Air Conditioning and Refrigeration Institute). Herbert Phillips, Vice President, Engineering.

Du Pont Chemicals. Dr. Helen A. Cannon, Manager Refrigerants Development Group; and John D. Burris, Senior Technical Assistant.

EPA (Environmental Protection Agency — Washington, D.C.). Debbie Ottinger and her staff.

EPA (Environmental Protection Agency — Dallas, TX). Bill Deese and Phyllis Putter.

UL (Underwriters Laboratories, Inc.). Larry Kettwich, Staff Engineer.

United Technologies, Carrier. James Parsnow, Product Marketing Manager, Centrifugal Chillers.

PART I

Refrigerants and Ozone Depletion

Chapter 1

Basic Theory of Ozone Depletion

Objectives

After completion of this chapter you should:

◆ *be more aware of the theory of ozone depletion.*

◆ *understand the effects of ozone depletion on human health.*

◆ *understand how ozone depletion effects plant and animal life.*

◆ *have a better understanding of the effect that ozone depletion has on the world.*

It has been almost two decades since scientists began to report that the ozone layer is being damaged. More studies have shown that this is true. Most of the damage is blamed on the chlorine atom in chlorofluorocarbon (CFC) and hydrochlorofluorocarbon (HCFC) refrigerants. The chlorine component in the chlorofluorocarbons (CFCs) and halons is causing the most damage. Laws have been enacted to prevent further damage to the ozone layer by placing controls on these refrigerants and causing more ozone-safe refrigerants to be developed and manufactured.

Ozone

Ozone (O_3) is an unstable, pale-blue gas with a penetrating odor; it is an allotropic form of oxygen usually formed by a silent electrical discharge

3

in the air. It forms a layer extending from a height of about 6 miles to 30 miles above the earth. In this layer there is an appreciable concentration of ozone, which absorbs ultraviolet radiation and prevents some loss of heat from the earth. Thus, if the ozone layer were not present, two things would occur: (1) ultraviolet, cancer causing rays would reach the earth, and (2) the earth would become warmer. Both of these occurrences have been taking place in the past few years. Skin cancer, cataracts, and suppression of the human immune system, all caused by ultraviolet rays, have increased, and the temperature of the earth has begun to rise in recent years. Both indicate that the ozone layer is being depleted.

There are two kinds of ozone with which we are familiar: stratospheric ozone and atmospheric ozone.

Stratospheric Ozone

The stratospheric ozone layer, also known as the ozone shield, is located at about 6 to 30 miles above the earth. It is this layer that absorbs and dissipates the ultraviolet rays from the sun, thus reducing the number of rays that reach the earth. The stratospheric ozone forms a protective layer surrounding the earth. It is the earth's main shield against ultraviolet rays from the sun. A decrease in stratospheric ozone allows more ultraviolet rays to reach the earth. In addition to health and earth-warming problems, an increase in ultraviolet rays presents a serious threat to our food crops by reducing crop yields. All forms of life on earth and in the sea are at risk.

Atmospheric Ozone

Ozone is also sometimes found closer to the earth. This is known as atmospheric ozone. As the ultraviolet rays from the sun reach the earth, they combine with smog and pollution and cause atmospheric ozone. This type of ozone is harmful and must not be confused with the stratospheric ozone layer. The oxygen in the air that we breathe is composed of two oxygen atoms (O_2). Ozone has three atoms (O_3). Thus, when there is a high concentration of the atmospheric ozone, we are breathing unhealthful air. On hot summer days when the atmospheric ozone is present in high concentrations we are warned to stay inside and not breathe any more of the ozone than is absolutely necessary. It is especially hard on persons having heart and respiratory problems.

Stratospheric Ozone Depletion

It has long been a theory that chemicals containing chlorine could damage the ozone layer. The problem is that chlorofluorocarbons (CFCs), hydrochlorofluorocarbons (HCFCs), and other similar compounds are persistent, extremely stable chemicals that rise up into the atmosphere intact until they reach the stratosphere. In the stratosphere, radiation from the sun breaks the molecule apart and sets free the chlorine component. Then the chlorine component attacks and destroys the ozone. One chlorine molecule can destroy approximately 1,000,000 ozone molecules.

In 1974, Dr. Sherwood Rowland and Dr. Marino Molina from the University of California published a paper demonstrating how CFCs destroy ozone in the stratosphere. At that time there were no measurements of actual ozone loss, just a scientific theory. Because of that theory, the United States led a ban on the use of CFCs in most aerosol applications.

More research in this area confirms the theory that CFCs continue to damage the ozone layer, and more is being learned about the way the ozone layer is damaged. Recent studies show a growing general agreement that we should reduce or completely eliminate CFCs. Environmentalists are concerned that the chlorine from these chemicals may cause further depletion of the ozone layer over regions where an increase in ultraviolet radiation can cause damage to living organisms. Industries throughout the world have joined together with governments and environmental groups to call for a revision of the control provisions of the Montreal Protocol to phase out CFCs, while at the same time making sure that safe replacements are available.

Because it is the chlorine in CFCs that can deplete the ozone, many scientists have used total atmospheric chlorine as a means of measuring the potential for ozone depletion. Although the relationship between the total atmospheric chlorine that reaches the ozone layer and the actual depletion is unknown at this time, any policy that reduces the amount of chlorine released into the atmosphere will provide additional long-term protection of the ozone layer.

The refrigerants that contain chlorine and not hydrogen are very stable chemicals. They are so stable that they do not break down readily in the lower atmosphere. It sometimes takes hundreds of years for them to chemically decompose. When released into the atmosphere, they gradually drift up into the stratosphere. The chlorine then reacts with the ozone, changing it back to oxygen. The problem is that the chlorine molecule is not changed; it continues to mix with other ozone molecules and cause more damage. This process continues for many years after the release of CFCs and halons into the atmosphere.

Halon fire extinguishers contain the chemical bromine. Bromine is chemically related to chlorine and acts the same way as CFCs in the ozone layer. However, halons are even more destructive than CFC refrigerants and have fallen into the same category.

Studies have been done to determine the effects that both CFC refrigerants and halons (used in fire extinguishers) have on the ozone layer. Each of these chemicals have been assigned a number that represents

5

their ozone-destroying ability, or ozone depletion potential (ODP). One study assigns CFC-11 an ODP factor of 1. All other refrigerants are rated as follows:

CFC and Halon	ODP
CFC-11	1.0
CFC-12	1.0
CFC-113	0.8
CFC-114	1.0
CFC-115	0.6
halon-1211	3.0
halon-1301	10.0
halon-2402	6.0

The EPA is banning these chemicals. Their schedule for elimination is in flux because of findings that the ozone layer continues to be damaged at a greater rate than previously believed. The effect of damage to the ozone layer on human and plant life is expanding.

Effects on Human Health

6

As mentioned earlier, the ozone layer protects the earth from the damaging effects of ultraviolet radiation from the sun. If more damage is done to the ozone layer, more of the damaging rays from the sun will reach the earth. It has been estimated that for each 1% depletion of the ozone layer approximately 1.5% to 2% increase in ultraviolet radiation will reach the earth. An increase in ultraviolet radiation will be accompanied by an increase in skin cancer, atmospheric ozone levels, global warming, eye cataracts, and damage to the human immune system.

The effects that ultraviolet radiation have on some of these conditions, such as skin cancer and cataracts, is common knowledge. Effects on the immune system are not completely understood.

Effects on Plant and Marine Life

Studies show that agricultural crops decrease with an increase in ultraviolet radiation. This research shows that some crops, when exposed to a concentration of ultraviolet radiation 25% higher than normal, had a decrease in yield of up to 20%.

Also, fish species that live close to the surface of the water will be exposed to more ultraviolet radiation. Researchers believe that an increase in ultraviolet radiation will cause a reduction in the productivity of these species of marine life.

Effects on the World

The CFC issue truly is a worldwide problem. What one country does greatly affects another. The release of CFCs and halons into the atmosphere and the resulting depletion of the ozone layer will affect the whole world. Every country that develops and uses CFCs and halons must make a commitment to curtail these activities. Approximate percentages of CFC consumption by different countries are:

United States	29%
former USSR and Eastern European nations	14%
developing nations	14%
other developed nations	41%
China and India	2%

Cooperation between worldwide policymakers and the CFC industry is necessary if we are to meet the needs of the society while protecting the ozone layer. Both requirements can be met with a rapid transition from CFCs to a combination of alternative products and technologies. Some of these alternatives include the use of new products that do not use CFCs or other fluorocarbons; increased conservation and cost-effective recovery; and the use of HFCs and HCFCs. Policymakers can facilitate an easier and more timely transition by establishing a realistic timetable for phasing out CFCs and halons, which is consistent with the availability of alternative products, and by committing themselves to long-term use of the alternatives.

7

Summary

◆ It has been almost two decades since scientists started reporting on ozone damage. Additional studies confirm this finding.

◆ Much of the damage has been blamed on the chlorine atom in chlorofluorocarbon (CFC) and hydrochlorofluorocarbon (HCFC) refrigerants.

◆ Chlorofluorocarbons and halons cause the most damage.

◆ Two kinds of ozone with which we are familiar are stratospheric ozone and atmospheric ozone.

◆ The stratospheric ozone layer absorbs and dissipates ultraviolet rays from the sun.

◆ A decrease in stratospheric ozone will allow more ultraviolet light to reach the earth.

◆ Atmospheric ozone is found closer to the earth. As the ultraviolet rays from the sun reach the earth, they combine with smog and pollution and cause atmospheric ozone. This is a harmful type of ozone, and it must not be confused with stratospheric ozone.

◆ Oxygen in the air that we breathe contains two oxygen atoms (O_2). Ozone has three atoms (O_3). When there is a high concentration of atmospheric ozone, we are breathing unhealthful air.

◆ On hot days, when atmospheric ozone is present in high concentrations, we are warned to stay inside and not breathe any more of the ozone than is absolutely necessary. It is especially hard on persons with heart and respiratory problems.

◆ Chlorofluorocarbons, hydrochlorofluorocarbons, and other compounds are persistent, stable chemicals that rise up into the atmosphere intact until they reach the stratosphere. In the stratosphere, radiation from the sun breaks the molecules apart and sets free the chlorine component, which destroys the ozone.

◆ Environmentalists are concerned that the chlorine from these chemicals may cause further depletion of the ozone layer over regions where increased ultraviolet radiation can damage living organisms.

8

◆ Industry leaders around the world have joined together with governments and environmental groups to call for a revision of the control provisions of the Montreal Protocol to phase out CFCs while making sure that safe replacements are available.

◆ The refrigerants that contain chlorine and not hydrogen are stable chemicals that may take hundreds of years to decompose chemically.

◆ Halon fire extinguishers contain bromine. Bromine is chemically related to chlorine and acts in the same way as CFCs in the ozone layer. Halons are more destructive than CFC refrigerants.

◆ For each 1% depletion of the ozone layer, approximately 1.5% to 2% more ultraviolet radiation will reach the earth.

◆ Studies have shown that agricultural crops are reduced with an increase in ultraviolet radiation.

◆ Species of fish that live close to the surface of the water will become exposed to more ultraviolet radiation.

◆ Cooperation between worldwide policymakers and all segments of the CFC industry is needed if we are to meet the needs of society while protecting the ozone layer. Both requirements can be met with a rapid transition from CFCs to a combination of alternative products and technologies.

◆ Key Terms ◆

atmospheric ozone
CFC (chlorofluorocarbon refrigerants)

HCFC (hydrochlorofluorocarbon refrigerants)
ODP (ozone depletion potential)
ozone (O_3)
stratospheric ozone
ultraviolet rays

Review Questions

Essay

1. What two chemicals cause the most damage to the ozone layer?
2. What is the earth's main shield against ultraviolet rays from the sun?
3. What are the two kinds of ozone with which we are familiar?
4. What is the harmful type of ozone?

True-False

5. Atmospheric ozone can cause problems for persons with heart and respiratory conditions.
6. The component in CFCs and HCFCs that destroys the ozone is oxygen.
7. Chlorofluorocarbon refrigerants are stable.
8. Bromine is placed in the same category as CFC and HCFC refrigerants.

9

Fill-in-the-Blank

9. The three refrigerants that have the greatest ozone depletion potential are _____, _____, and _____.
10. Each 1% ozone depletion will allow ___% to ___% increase in ultraviolet radiation to reach the earth.
11. For each 25% increase in ultraviolet radiation that reaches the earth, there is a _____% reduction in agricultural crops.
12. Some species of _____ will be adversely affected by depletion of the ozone layer.

Multiple Choice

13. Ozone is an unstable, pale-blue gas, with a penetrating odor; it is a form of oxygen formed by
    a. the joining of wind from two different directions.
    b. rain clouds.
    c. a silent electrical discharge in the air.
    d. the combining of water and air.

14. The most destructive chemical to the ozone is
    a. CFC.
    b. HCFC.
    c. bromine.
    d. chlorine.
15. The CFC problem is felt
    a. worldwide.
    b. nationwide.
    c. statewide.
    d. citywide.

10

CFCs: Their Problems and Alternatives

Objectives

After completion of this chapter you should:

◆ *know more about the history of refrigerants.*

◆ *know more about the types of refrigerants being banned.*

◆ *know why these specific refrigerants are being banned.*

◆ *know more about the leading candidates for the replacement of CFCs.*

◆ *be more familiar with the safety and performance of the alternative refrigerants.*

◆ *understand more of the problems of compressor lubrication with alternative refrigerants.*

◆ *be more familiar with the requirements of refrigerant cylinders.*

Since the beginning of time people have sought some means of cooling their foods and perishables. This was first done with water. Then experiments and necessity led to other means for cooling.

History of Refrigerants

Through the ages, people have used some means of cooling foods and themselves. At first, they cooled foods by lowering them into wells or storing them in caves. Then natural ice was used. Ice was cut in the winter and stored underground for use during the warm months.

The Egyptians found that they could keep food cool by storing it in clay containers. This method was quite successful because the water slowly seeped through the porous walls and dried on the outside wall. This drying action, known as evaporation, caused the jar and its contents to cool. The principle of evaporation of a liquid is the basis of modern mechanical refrigeration.

The old fashioned icebox came into service with the use of natural ice. The problem of transportation was a difficult one. Consequently, natural ice was found to be a very expensive luxury in warm, dry areas.

More than one hundred years ago, an English scientist succeeded in changing ammonia gas to a liquid by applying pressure and lowering the temperature. When the pressure was released, the ammonia boiled off rapidly and changed back into a gas. When this happened he found that heat was removed from the surroundings. This discovery proved to be of tremendous importance in the development of the modern mechanical refrigeration unit.

The first ice machine appeared in 1825. The ice produced was purer and cleaner than natural ice and was independent of weather conditions.

12

Types of Refrigerants

From this point, the refrigerant family grew and became more specialized and complicated. Some refrigerants were found to have undesirable characteristics and were discarded for the more desirable types.

The modern refrigerants are generally divided into three specific categories. These categories are chlorofluorocarbons, hydrochlorofluorocarbons, and hydrofluorocarbon.

Chlorofluorocarbons (CFCs) are some of the most useful chemical compounds ever developed. They help meet the needs for most goods and services related to food, shelter, health care, energy efficiency, communications, and transportation. They also contribute immensely to human safety, productivity, and comfort.

These refrigerants were developed almost 60 years ago. They are nonflammable, noncorrosive, not highly toxic, and are compatible with many types of materials. Chlorofluorocarbons have the thermodynamic and physical properties that make them useful for many applications. They are used in blowing agents in the manufacture of insulation, refrigerants, packaging and cushioning foams, cleaning agents for metal and electronic components, and in many other common operations.

However, chlorofluorocarbon refrigerants are reported to be causing most of the damage to the ozone layer. The ozone layer is in the atmosphere from just a few to several miles above the earth. The ozone layer protects the earth from the sun's ultraviolet solar radiation rays.

The following list shows the designation, chemical formula, and the molecular structure of the CFC refrigerants that are being controlled by a public law created by the Montreal Protocol and governed by the United States Environmental Protection Agency (EPA). These regulations were adopted July 1, 1989, in Montreal, Quebec, Canada.

R-11	Trichlorofluoromethane	CCl_3F
R-12	Dichlorodifluoromethane	CCl_2F_2
R-113	Trichlorotrifluoroethane	CCl_2FCClF_2
R-114	Dichlorotetrafluoroethane	$CClF_2CClF_2$
R-115	Chloropentafluorethane	$CClF_2CF_3$

These types of refrigerants consist of chlorine, fluorine, and carbon. They have no hydrogen in their composition and are, therefore, very stable chemically. They tend to remain unchanged for many years after being released into the atmosphere. Because they are stable and have the chlorine component, these refrigerants have a very high ozone depletion potential (ODP). The president of the United States has asked that their production be stopped by July 1, 1995.

13

Alternatives to Chlorofluorocarbons

The leading candidates for the replacement of chlorofluorocarbon refrigerants are hydrofluorocarbon and hydrochlorofluorocarbon refrigerants.

The hydrofluorocarbon refrigerants do not contain chlorine. Therefore, they have a zero ozone depletion potential. In hydrochlorofluorocarbon refrigerants, the hydrogen (H) added to the chemical makeup of CFCs allows the dissipation of virtually all the chlorine in the lower part of the atmosphere, before it has a chance to reach the ozone layer. Because the chlorine dissipates at lower altitudes, HCFCs have a much lower ODP, from 2% to 10% of that of CFC refrigerants. The hydrogen in both HFC refrigerants and HCFC refrigerants causes them to be less stable in the atmosphere than are the CFC refrigerants. They also have greatly reduced atmospheric lifetimes, from 2 to 25 years, compared to about 100 years or longer for the CFC refrigerants.

In this book we discuss nine alternatives for CFC refrigerants: four HFCs (HFC-23, HFC-125, HFC-134a, HFC-152a), and five HCFCs (HCFC-22, HCFC-123, HCFC-124, HCFC-141b, HCFC-142b). The molecular weight, chemical formula, and boiling point for each of these refrigerants are given in Table 2-1.

Product	Formula	Molecular Weight	Boiling Point	
			°C	°F
HFC-23	CHF_3	70.01	−82.03	−115.66
HFC-125	CHF_2CF_3	120.02	−48.50	−55.30
HCFC-22	$CHClF_2$	86.47	−40.75	−41.36
HFC-134a	CH_2FCF_3	102.00	−26.50	−15.70
HFC-152a	CH_3CHF_2	66.00	−24.70	−12.50
HCFC-124	$CHClFCF_3$	136.50	−11.00	12.20
HCFC-142b	CH_3CClF_2	100.47	−9.80	14.40
HCFC-123	CF_3CHCl_2	152.90	27.90	82.20
HCFC-141b	CH_3CCl_2F	116.95	32.00	89.60

Table 2-1   Alternatives to CFCs. *Courtesy of Du Pont Chemicals.*

Safety and Performance of the Alternatives

Testing has shown that HFCs and HCFCs exhibit properties and performance characteristics similar to CFCs, but they do not have the environmental impact of the CFCs. HFCs and HCFCs have a low toxic rate, are stable, have either no or low flammability characteristics, produce no smog, and require minimal changes to equipment when compared to other not-in-kind types of refrigerants.

A joint toxicity effort, the Program for Alternative Fluorocarbon Toxicity Testing (PAFTT), is under way to accelerate toxicity testing on alternatives. The final toxicity results for the refrigerants HFC-134a, HCFC-123, and HCFC-141b should be available from PAFTT in 1993. The final toxicity results for HCFC-124 and HFC-125, also being tested by PAFTT, should be available in 1994-1995. The other alternative refrigerants shown here (HCFC-22, HFC-23, HFC-152a, and HCFC-142b) are low in toxicity and have been used commercially for many years in applications such as air conditioning and aerosol propellants. The current toxicity data reflecting workplace safety standards (TLVs or AELs) is shown in Table 2-2, which shows the physical properties of HFCs and HCFCs.

Figure 2-1 shows that HCFCs and HFCs provide great improvements in terms of both ozone depletion potential and halocarbon global warming potential (GWP).

The area of each circle is proportional to the atmospheric lifetime of the refrigerant that it represents. The center of the circle marks the ODP and the halocarbon GWP. Included in this grouping is carbon tetrachloride (CTC).

Figure 2-2 represents another factor by which alternatives must be judged: the potential to contribute to smog formation.

Additional compounds shown here are the hydrocarbons (methane, ethane, propane, butane, pentane), dimethyl ether (DME), methylene chloride (MC), and methyl chloroform (MeClf). The EPA has determined that the compounds above the photochemical reactivity line (the diago-

14

(a) Denotes calculated values.

(b) Values calculated using the Du Pont Model and best available information. CFC-11 is the reference compound for both ozone depletion potential and halocarbon global warming potential.

(c) CFC-12 has a halocarbon global warming potential of 3. Thus, to determine the gain in switching from CFC-12 to one of the alternatives, the value shown in the table should be divided by 3.

(d) TLV (Threshold Limit Value) is a registered trademark of the American Conference of Governmental Industrial Hygienists. AEL (Allowable Exposure Limit) is a preliminary toxicity assessment established by Du Pont, additional studies may be required.

(e) National Institute of Standards and Technology (NIST).

		HCFC-22	HFC-23	HCFC-123	HCFC-124	HFC-125	HFC-134a	HCFC-141b	HCFC-142b	HFC-152a
Chemical Name		Methane, Chlorodifluoro	Methane, Trifluoro	Ethane, 2,2-Dichloro, 1,1,1-Trifluoro	Ethane, 2-Chloro-1,1,1,2-Tetrafluoro	Ethane, Pentafluoro	Ethane, 1,1,1,2-Tetrafluoro	Ethane, 1,1-Dichloro-1-Fluoro	Ethane, 1-Chloro, -1,1-Difluoro	Ethane, 1,1-Difluoro
Chemical Formula		$CHClF_2$	CHF_3	$CHCl_2CF_3$	$CHClFCF_3$	CHF_2CF_3	CH_2FCF_3	CH_3CCl_2F	CH_3CClF_2	CH_3CHF_2
Molecular Weight		86.47	70.01	152.9	136.5	120.02	102.0	116.95	100.47	66.0
Boiling Point at 1 atm	°C / °F	−40.75 / −41.36	−82.03 / −115.66	27.9 / 82.2	−11.0 / 12.2	−48.5 / −55.3	−26.5 / −15.7	32.0 / 89.6	−9.8 / 14.4	−24.7 / −12.5
Freezing Point	°C / °F	−160.0 / −256.0	−155.2 / −247.4	−107.0 / −161.0	−199.0 / −326.0	−103.0 / −153.0	−101.0 / −149.8	−103.5 / −154.3	−130.8 / −203.4	−117.0 / −179.0
Critical Temperature	°C / °F	96.0 / 204.8	25.9 / 78.6	185.0 / 365.0	122.2 / 252.0	66.3 / 151.3	100.6 / 213.9(a)	210.3 / 410.5	137.1 / 278.8	113.5 / 236.3
Critical Pressure	atm / lbs/sq in abs	49.12 / 721.9	47.7 / 701.4	37.4 / 549.6	35.27 / 518.3	34.72 / 510.3	40.03 / 588.3(a)	45.80 / 673.1	40.7 / 598.0	44.4 / 652.5
Critical Volume	cc/mol / cu ft/lb	165 / 0.0305	133 / 0.0305	288.8(a) / 0.0302	246.4 / 0.0289	210.2 / 0.0281	208.8 / 0.0311(a)	270.0(a) / 0.0370	231.0 / 0.0368	na / na
Critical Density	g/cc / lbs/cu ft	0.525 / 32.76	0.525 / 32.78	0.53 / 33.11	0.554 / 34.61	0.571 / 35.68	0.515 / 32.2(a)	0.43(a) / 26.87	0.435 / 27.18	0.365 / 22.81
Density, Liquid @ 25°C (77°F)	g/cc / lbs/cu ft	1.194 / 74.53	0.670 / 41.82	1.46 / 91.15	1.364 / 85.15	1.25 @ 20°C / 78.0	1.202 / 75.10	1.23 / 76.70	1.12 @ 20°C / 69.90	0.911 / 56.90
Density, Sat'd Vapor @ Boiling Point	g/l / lbs/cu ft	4.72 / 0.295	4.66 / 0.291	5.8 / 0.362	6.882 / 0.430	6.56 / 0.410	5.04 / 0.315	4.67 / 0.292	4.72 / 0.295	na / na
Specific Heat, Liquid (Heat Capacity) @ 25°C (77°F)	cal/g(°C) or Btu/(lb)(°F)	0.300	0.345 @ −30°C (−22°F)	0.243	0.270	0.301(a)	0.341	0.276	0.310	0.400
Specific Heat, Vapor, at Const Pressure (1 atm) @ 25°C (77°F)	cal/g(°C) or Btu/(lb)(°F)	0.157	0.176	0.172	0.177	0.169(a)	0.205	0.189	0.210	0.280
Vapor Pressure (psia) @ 25°C (77°F)		150	660	14	61	190	96	11	49	87
Heat of Vaporization @ Boiling Point	cal/g / Btu/lb	55.81 / 100.45	57.23 / 103.02	41.6 / 74.9	40.1(a) / 72.2	38.0 / 68.4	52.4(a) / 94.5	53.3 / 95.9	53.3 / 95.9	53.25 / 95.85
Thermal Conductivity @ 25°C (77°F) Btu/(hr)(ft)(°F) Liquid / Vapor (1 atm)		0.0507 / 0.00609	(@ −30°C) 0.0569 / 0.0060	0.0471 / 0.0055	0.0417 / 0.0075	0.0363 / 0.0084	0.0474 / 0.0084	0.0546 / 0.0050	0.0480 / 0.0075	0.0600 / 0.0085
Viscosity @ 25°C (77°F) Liquid / Vapor (1 atm)	centipoise	0.198 / 0.0127	(@ 30°C) 0.167 / 0.0118	0.449 / 0.0130	0.314 / 0.0131	0.104(a) / 0.015(a)	0.205 / 0.0137	0.409 / 0.0125	0.320 / na	0.239 / na
Solubility of HFC or HCFC in Water @ 1 atm and 25°C (77°F)	wt%	0.30	0.10	0.39	0.145	0.09	0.15	0.021	0.14	0.28
Solubility of Water in HFC or HCFC @ 25°C (77°F)	wt%	0.13		0.08	0.07	0.07	0.11	na	0.05	0.17
Flammability Limits in Air	vol%	None	None	None	None	None	None	7.3–16.0	6.7–14.9	3.9–16.9
Ozone Depletion Potential(b)		0.05	0	0.02	0.02	0	0	0.10	0.06	0
Halocarbon Global Warming Potential(c)		0.3		0.02	0.10	0.58	0.26	0.09	0.36	0.03
TSCA Inventory Status		Reported/Included	Reported/Included	Reported/Included	Not in Inventory	Not in Inventory	Reported/Included	Not in Inventory	Reported/Included	Reported/Included
Toxicity TLV or AEL(d)	ppm (v/v)	1,000 (TLV)	1,000 (AEL)	100 (AEL)	500 (AEL)	1,000 (AEL)	1,000 (AEL)	100 (AEL)	1,000 (AEL)	1,000 (AEL)

Table 2-2   Physical properties of HFCs and HCFCs. *Courtesy of Du Pont Chemicals.*

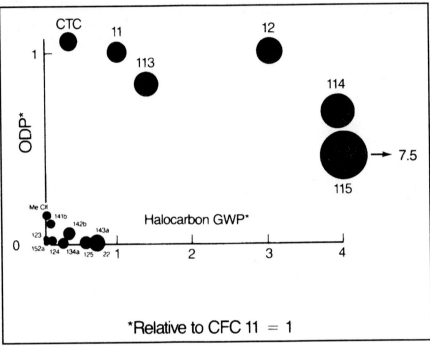

Figure 2-1   GWP of several refrigerants relative to CFC-11. *Courtesy of Du Pont Chemicals.*

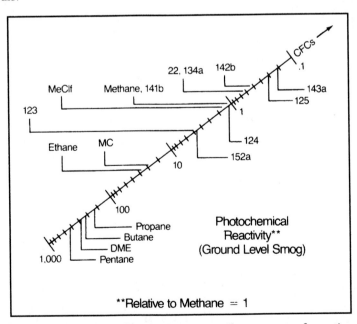

Figure 2-2   Potential of CFC alternatives to contribute to smog formation. *Courtesy of Du Pont Chemicals.*

nal line) are negligibly photochemically reactive and has exempted them from the U.S. volatile organic compound (VOC) regulations. The HCFCs and the HFCs below the photochemical reactivity line are under consideration for exemption.

Applications of HFCs and HCFCs is shown in Table 2-3. It is not an all-inclusive list. For specialized applications or more detail make a special request to one of the refrigerant manufacturers.

HFC or HCFC Compound	Refrigerants Applications	Blowing Agents Applications	Other Uses
HFC-23	Possible low temperature alternative.		Plasma etchant.
HFC-125	Blend component for low temperature and medium temperature applications.		
HCFC-22	Low temperature and medium temperature applications; household, commercial refrigeration and air conditioning applications. Component in ternary refrigerant blend.	Polystyrene, polyethylene, polyurethane, polyisocyanurate and phenolic	Propellant.
HFC-134a	Household and commercial refrigeration systems, automotive air conditioning. Medium temperature food cases and chillers.	Polystyrene, polyurethane, polyisocyanurate and phenolic	Propellant for aerosol drugs/pharmaceuticals.
HFC-152a	Component in ternary blend for medium temperature applications and automotive air conditioning service.	Polystyrene, polyurethane, polyisocyanurate and phenolic	Propellant.
HCFC-124	Chillers. Component in ternary blend for medium temperature applications and automotive air conditioning service.	Polystyrene, polyethylene, polyurethane, polyisocyanurate and phenolic	
HCFC-142b		Polystyrene, polyurethane, polyethylene, polyisocyanurate and phenolic	Propellant.
HCFC-123	Chillers.	Polyurethane, polystyrene, polyisocyanurate and phenolic	Component in cleaning agent blend.
HCFC-141b		Polyurethane, polyisocyanurate and phenolic	Component in cleaning agent blend.

17

Table 2-3   Applications of HFC and HCFC compounds. *Courtesy of Du Pont Chemicals.*

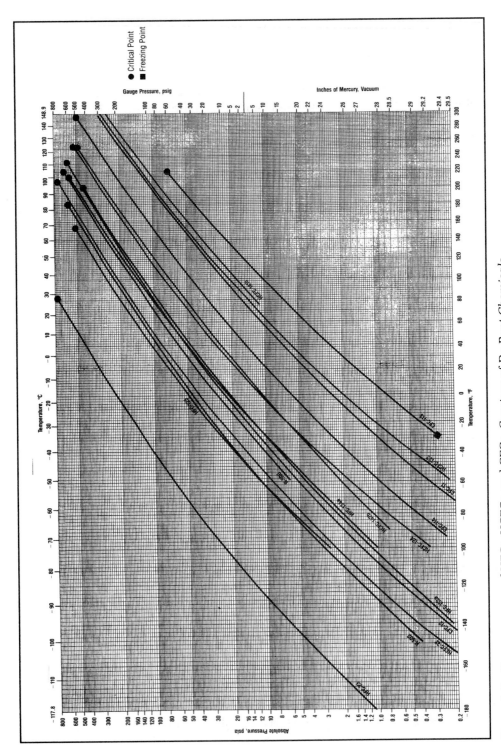

Figure 2-3   Vapor pressure of HFCs, HCFCs, and CFCs. *Courtesy of Du Pont Chemicals.*

Applications	Current	Alternative
Appliances	CFC-12	HFC-134a
	R-500	Blends
Chillers	CFC-11	HCFC-123
	CFC-12	HFC-134a
	CFC-114	HCFC-124
Retail food		
Low temperature	R-502	HCFC-22
		HFC-125
Medium temperature	R-502	HCFC-22
	HCFC-22	HFC-125
	CFC-12	HFC-134a
		Blends
Mobile A/C	CFC-12	HFC-134a
		Blends

Table 2-4   Alternative refrigerant applications. *Courtesy of Du Pont Chemicals.*

19

Figure 2-3 presents the vapor pressure of HFCs, HCFCs, and CFCs. See Table 2-2, which lists the physical properties of HFCs and HCFCs.

With the phaseout of the CFCs and HCFCs, refrigerant manufacturers are conducting research to develop new refrigerants. Table 2-4 lists the applications of alternative refrigerants, the current CFC used, and the alternative refrigerant.

A comparison of vapor pressure of existing refrigerants and their alternatives is shown in Figure 2-4. Figure 2-5 shows the relationship between the high performance blends for the CFC-12 aftermarket.

These blends may be used for this market.

Figure 2-6 shows a comparison of the vapor pressures of the replacement blends compared to CFC-12 vapor pressures for the CFC-12 aftermarket.

Table 2-5 shows the theoretical cooling capacity and energy efficiency improvements of the replacement blends compared to CFC-12 being replaced.

Oils for Alternative Refrigerants

An important consideration in the development of hydrochlorofluorocarbons and hydrofluorocarbons as alternative refrigerants is the testing

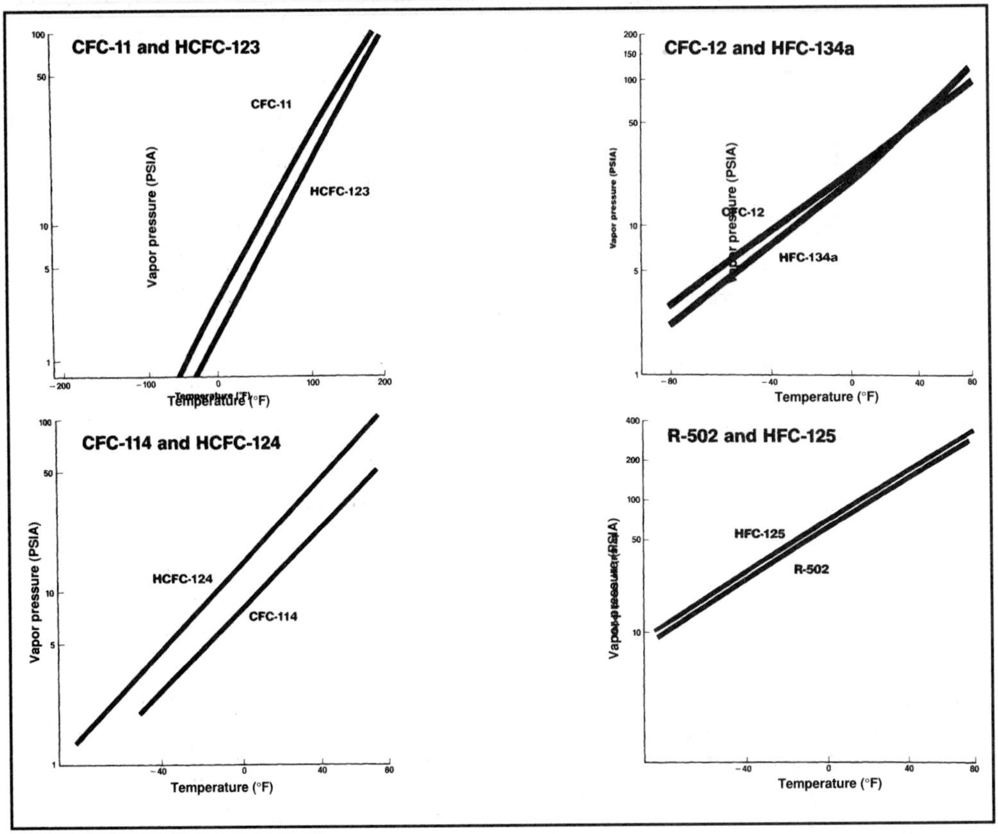

Figure 2-4   Vapor pressure comparisons of existing refrigerants and alternatives. *Courtesy of Du Pont Chemicals.*

of oils for solubility, stability, lubricity with the refrigerant, and compatibility with materials of construction. Testing will determine which lubricants can be used in current systems and what changes are required.

In general the ternary blends and HCFCs can be used with oils that are commercially available; however, some development work may be required to optimize performance. Applications with HFCs require the development of new oils. The desirable properties for developmental use are:

1. acceptable solubility with the refrigerant (ideally, single phase over a broad temperature range);
2. acceptable lubricity;
3. good thermal stability for the refrigerant/lubricant combination;
4. acceptable compatibility with system materials (elastomers, metals, and plastics);
5. low toxicity; and
6. commercial availability at a reasonable cost.

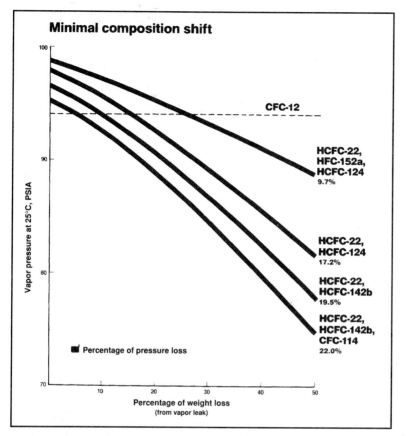

Figure 2-5   High performance blends for the CFC-12 aftermarket. *Courtesy of Du Pont Chemicals.*

21

Test results for nonproprietary oils with ternary blends and proprietary oils with HFC-134a are included here. Tests for other proprietary oils and refrigerant/lubricant combinations are currently under way or being planned.

HFC-134a

HFC-134a is the alternative choice for CFC-12 in automotive air conditioning and many stationary refrigeration/air conditioning applications. Extensive testing is under way in various industries to evaluate the performance of HFC-134a and determine what design changes will be necessary to optimize its performance.

Solubility. Existing refrigeration/air conditioning mineral oils are completely soluble in CFC-12 over a wide temperature range. This ensures

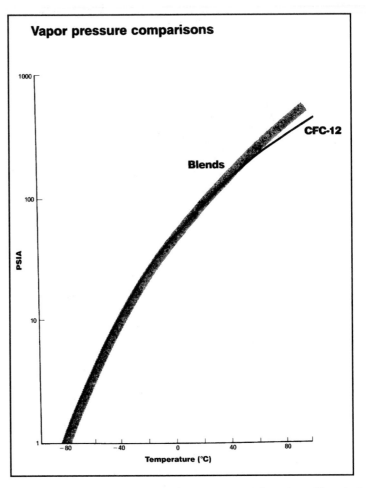

Figure 2-6   Blends for CFC-12 aftermarket. *Courtesy of Du Pont Chemicals.*

22

that the oil moves freely around the system and returns to the compressor at a rate sufficient to provide acceptable lubrication. Many of the polyalkylene glycols (PAGs) being evaluated for use in these systems are not completely soluble with HFC-134a. Several examples are shown in Figure 2-7 where the PAGs show completely different solubility characteristics in HFC-134a as compared to CFC-12/mineral oils.

CFC-12 mineral oils separate into two phases below their solubility curves, whereas the HFC-134a/PAGs separate into two phases above the curves. This means that in the parts of the system that operate above the minimum solubility temperature and at certain oil/refrigerant ratios, two phases will occur and the oil-rich portion might collect at these locations. This could cause flow restrictions, reduced heat transfer, and possibly insufficient oil return to the compressor. Individual system

	CFC-12	Blend (range)
Cooling capacity (relative CFC-12)	1.0	0.95–1.05
Energy efficiency (COP)	2.10	2.16–2.17
Evaporator pressure (PSIA)	19	16–18
Compressor discharge pressure (PSIA)	196	195–215
Compressor discharge temperature (°F)	158	185–190

Test conditions: Condenser 130°F, Sub-cooling 0, Evaporator −10°F, Superheat 5°F

- Patented multi-component blends, can contain: HCFC-22, HFC-152a, CFC-114, HCFC-124 and others
- Soluble with commercial oils
- Minimal retrofit expected
- Applications may include medium temperature food cases, vending machines, auto a/c and appliances
- Possible OEM candidates

23

Table 2-5   Theoretical cooling capacity and energy efficiency improvement using blends. *Courtesy of Du Pont Chemicals.*

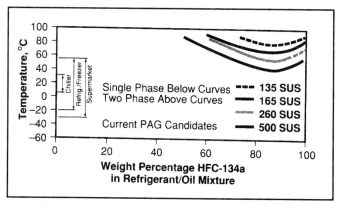

Figure 2-7   Candidate oils solubility with HFC-134a. *Courtesy of Du Pont Chemicals.*

85/15 Volume % Refrigerant/Oil Solution Tested at 131°C for 11.8 Days						
				Effect Rating		
Refrigerant	**Oil**	**Halides, ppm**	**Color**	**Fe**	**Cu**	**Al**
HFC-134a	PAG	<0.2	0	0	0	0
CFC-12	Naphthenic	423	4	3	2	2
CFC-12	Paraffinic	—	0	3	0	0

90/10 Volume % Refrigerant/Oil Solution Tested at 177°C for 3 Days						
				Effect Rating		
Refrigerant	**Oil**	**Halides, ppm**	**Color**	**Fe**	**Cu**	**Al**
HFC-134a	PAG	<0.2	0	2	0	0
HFC-134a (with nylon)	PAG	<0.2	0	0	0	0
CFC-12	Paraffinic	611	1	2	0	0

Rating Range: 0 = no effect to 5 = severe effect.

24

Table 2-6   Stability tests for HFC-134a in various oils. *Courtesy of Du Pont Chemicals.*

testing must be conducted to determine the impact that partial solubility might have on system performance and compressor durability.

Chemical Stability. Sealed-tube tests of HFC-134a/PAG combinations have been run to determine their stability when exposed to copper, steel, and aluminum. Table 2-6 summarizes the stability test results of mixtures with 85% and 90% volume of refrigerant.

These test results indicate that HFC-134a/PAG solutions have acceptable chemical stability. Other tests have confirmed that the HFC-134a molecule is at least as stable chemically as CFC-12.

Additional HFC-134a/Lubricant Testing Required. Because PAGs are commercially available oils, the refrigerant manufacturers are working with companies worldwide to evaluate them as possible lubricants with HFCs for a wide variety of refrigeration and air conditioning applications. The evaluation covers a viscosity range of approximately 100 SSU to 700 SSU (Saybolt Seconds Universal) and is intended to determine:

1. solubility of PAG in HFC;
2. lubricity of the HFC/PAG combination; and
3. compatibility of the HFC/PAG combination with materials such as "O"-rings, gaskets, desiccants, metals, and plastics.

Based on testing to date, several issues about the use of PAGs with HFC-134a have been identified and will require further evaluation, including:

1. the long-term impact of complete solubility on compressor durability;
2. the role of PAG in copper plating;
3. the role of PAG in hermetic motor materials incompatibility;
4. the effect of the hygroscopic nature of PAGs in creating unacceptable moisture levels in the system; and
5. the fact that certain PAGs exhibit less than desirable lubricity for steel-on-aluminum surfaces.

Similar, small-scale evaluations are being conducted on non-PAG oils with improved properties. However, these lubricants are in early development and are not available for widespread testing. As more knowledge is gained, they will be made available in limited quantities for testing specific applications.

Ternary Refrigerant Blends

25

Ternary refrigerant blends are considered to be strong alternative candidates for CFC-12 in certain refrigeration/air conditioning applications, including automotive air conditioning aftermarket, refrigerators and freezers, and other stationary applications. Two blends, KCD-9430 and KCD-9433, were used for the oil testing; their compositions (by weight) were as follows:

KCD-9430: 36% HCFC-22; 24% HFC-152a; 40% CFC-114
KCD-9433: 36% HCFC-22; 24% HFC-152a; 40% HCFC-124

Oil Solubility/Lubricity. Oil solubility tests were run for 30, 60, and 90 weight percent of the ternary blends in 500 SSU viscosity paraffinic, naphthenic, and alkylbenzene oils across a temperature range of −10°C to 93°C. The test goal was complete solubility in all refrigeration/oil concentrations throughout this temperature range. The paraffinic and naphthenic oils did not meet the solubility goal; however, the alkylbenzene oil was soluble over the complete range of concentrations and temperatures.

Lubricity tests show that ternary blends/alkylbenzene mixes do not perform without additives. However, it has been demonstrated that mixtures of the blends and alkylbenzene lubricants respond well to the type of extreme pressure additives that have been used in refrigeration and air conditioning systems for many years.

Chemical Stability. Sealed-tube stability tests were performed on the ternary blends with four oils (naphthenic, alkylbenzene, and two paraffinic

| Oil/Refrigerant | Effect Rating | | | Cl⁻, ppm | F⁻, ppm |
	Liquid Color	Metals	Copper Plating		
Parafinnic Oil A					
KCD-9430	3	1	2	11.1	3.6
KCD-9433	1	0	0	0.7	0
CFC-12	5	2	2	20.3	6.3
Paraffinic Oil B					
KCD-9430	0	1	1	3.2	0
KCD-9433	0	0	0	0.02	0
CFC-12	0	1	0	2.0	0
Naphthenic Oil					
KCD-9430	3	0.5	2	17.4	6.1
KCD-9433	2	0.5	0.5	0.06	0
CFC-12	8	2	2	5.4	5.5
Alkylbenzene Oil					
KCD-9430	1	1	1	7.2	1.9
KCD-9433	1	0	0	0.02	0
CFC-12	2	2	2	6.9	0

Codes:
  Liquid: 0=Colorless, 1=Very light yellow, 2=Light yellow,
          3=Yellow, 5=Light tan, 8=Black
  Metals: 0=No change, 1=Slight, 2=Medium, 3=Heavy
  Copper Plating: 0=None, 1=Light, 2=Heavy

Table 2-7 Chemical stability tests for ternary blends vs. CFC-12. *Courtesy of Du Pont Chemicals.*

oils) in contact with copper, steel, and aluminum. The tests were conducted at 150°C for 30 days, with analysis of color development, effect on metals, and chloride/fluoride content. Table 2-7 shows these results.

The stability of ternary blend/alkylbenzene combinations is similar to that of the most stable mineral oil with the blends, and more stable than the existing CFC-12/mineral oil systems. Also, the blend containing HCFC-124 (KCD-9433) is more stable than the CFC-114 blend (KCD-9430).

Summary of Oils for Alternative Refrigerants

Both ternary blends and HCFCs can be used with oils that are commercially available; however, some system changes or lubricant modifications may be required to optimize performance. Evaluations on existing oils are being conducted, and work continues on the development of lubricants.

Testing to date indicates that HFC-134a/PAG combinations have acceptable chemical stability. However, several issues concerning the use of PAGs with HFC-134a, such as the long-term impact of incomplete solubility on compressor durability, require further evaluation. In addition, tests are being conducted on several non-PAG oils with improved properties to determine their potential use with HFC-134a.

Oil solubility tests conducted over 30, 60, and 90 weight percent of the ternary blends show that alkylbenzene oil is soluble from −10°C to 93°C over the complete range of concentrations. Lubricity tests show that ternary blend/alkylbenzene combinations do not perform well without additives. On the other hand, they respond well to the types of extreme pressure additives that are used in many refrigeration/air conditioning systems. The chemical stability of the ternary blends/alkylbenzene combinations is better than that of existing CFC-12/mineral oil systems.

Although existing mineral oils are not completely soluble in the ternary blends, there is still a possibility that they can be used. Additional compressor durability tests are required.

Refrigerant Cylinders

27

Refrigerant cylinders are manufactured in a variety of types and sizes. The types are disposable and returnable. The returnable cylinders require a deposit at the time of purchase. The deposit is refunded at the time the cylinder is returned or exchanged for a full one. They come in sizes ranging from 15 pounds to 125 pounds. Most of the 15-, 30-, and 50-pound cylinders are disposable. The disposable type is the most popular because no deposit is necessary, and disposable types are generally easier to handle. The type of refrigerant in the cylinder can be determined by the color of the cylinder. The universal color code is:

R-11	orange
R-12	white
R-22	green
R-500	yellow
R-502	purple
recovery cylinder	yellow top with a gray body

Cylinder Regulations

All refrigerant cylinders are manufactured to meet specifications established by the U. S. Department of Transportation (DOT). This agency has the authority to regulate all hazardous materials shipped through commercial means. CFC disposable cylinders are manufactured to meet DOT

Specification 39. Disposable cylinders are, therefore, called DOT 39s. These cylinders must be tested and meet the requirements for the highest pressure refrigerant encountered, which is R-502. This calls for a service pressure of 260 psi (pounds per square inch) test pressure. The test pressure for disposable cylinders is 350 psi and they must not rupture below 650 psi.

Disposable cylinders are manufactured from steel products and are therefore subject to rusting. Thus, they must be stored in a dry place. If the cylinder walls become rusty, they may not be able to hold the pressures required of them. From this it can be seen that the use of disposable cylinders for compressed air tanks is hazardous. The cylinder will rust from the inside where the damage is not noticeable. When the cylinder is subjected to compressed air pressures it may rupture and injure anyone standing in the vicinity.

It is illegal to transport a DOT 39 cylinder that has been refilled. This illegal act has a maximum penalty of $25,000 and five years in prison. Also, the refilling of DOT 39 cylinders is a violation of many state and local laws.

Disposable cylinders should be disposed of properly. Their contents can be removed with a recovery unit; the cylinder then is recycled with scrap metal. Never purge a cylinder to the atmosphere. Never leave a partially filled cylinder laying around. Someone accidentally could puncture the cylinder and cause serious injury.

Recovery cylinders must comply with DOT specifications. These are special cylinders designated for transporting contaminated refrigerants or refrigerant that has been recovered from a system and taken from the job site.

Recovery cylinders are painted yellow on the top and 12 inches down the side. The bottom of the cylinder is painted gray. Every cylinder has "MAX. GROSS WT._____LB." stenciled on the shoulder. This is the gross weight of the filled cylinder. It must never be exceeded. When recovering refrigerant from a system, use only recovery cylinders. Never use cylinders that are designed for new, or virgin, refrigerant.

28

Safety

The following is a discussion of safety for both the refrigerant cylinders and health safety. These suggestions are not intended to replace any safety procedures recommended by the equipment manufacturer, the cylinder manufacturer, or the refrigerant manufacturer.

Refrigerant Cylinders. The DOT requires that every cylinder be manufactured with a safety relief that will relieve pressure from inside the cylinder before the cylinder reaches the bursting pressure. There are two types of safety relief devices designed for DOT 39 cylinders. They are the frangible disc and a spring loaded relief valve. The frangible disc type is welded to the shoulder of the cylinder. When the pressure inside the cylinder exceeds 340 psi, the disc will rupture and release the refrigerant before the cylinder ruptures. The spring loaded device is preset to lift off its seat if the pressure inside the cylinder should exceed 340 psi. The pressure inside the cylinder forces the valve off its seat to release the pressure before the pressure inside the cylinder reaches the cylinder bursting point. The cylinder contents are released through a relief port.

There are several reasons for a refrigerant cylinder to become overpressurized. Some of the major reasons are too much liquid refrigerant inside; storage in a hot area, such as in direct sunlight; and overpressurizing from the compressor discharge.

When filling a recovery cylinder, make certain that it is not a DOT 39 type. Refilling these types of cylinders is illegal. Prior to filling a cylinder, be sure to check it for any external damage. If any external damage is present, do not refill the cylinder.

ARI Guideline K requires the following filling procedure: IMPORTANT: DO NOT MIX REFRIGERANTS WHEN FILLING CONTAINERS.

Cylinders and Ton Tanks: Do not fill if the present date is more than 5 years past the test date on the container. The test date will be stamped on the shoulder or collar of cylinders and on the valve end chime of ton tanks and appear as follows:

29

A1

12   89

32

This indicates the cylinder was retested in December of 1989 by retester number A123.

Cylinders and ton tanks shall be weighed during filling to ensure user safety. "MAXIMUM GROSS WEIGHT" is indicated on the side of the cylinder or ton tank and shall never be exceeded.

Cylinders and ton tanks shall be checked for leakage prior to shipment. Leaking cylinders must not be shipped.

Never fill a cylinder more than 80% of its rated weight capacity. When an overfilled refrigerant cylinder is exposed to higher temperatures the pressure inside will increase very rapidly. If the pressure rises too fast for the safety device to sense it, the cylinder may explode causing damage and possible personal injury.

When a refrigerant cylinder is to be subjected to temperatures of more than 130°F, the cylinder should only be filled to

60% capacity because the refrigerant expands to fill the cylinder with liquid. This is known as a hydrostatic condition. At this point the pressure inside increases rapidly with only a small change in temperature. Should the safety device be unable to detect this sudden increase in pressure, the tank probably would rupture causing damage, personal injury, or even death.

Never store a refrigerant cylinder in direct sunlight. The rays from the sun will heat the cylinder, causing a rapid increase in pressure. If the temperature reaches 130°F the pressure will increase very rapidly.

The following list should be followed in addition to any safety procedures recommended by the equipment or cylinder manufacturer:

1. Never mix refrigerants.
2. Refill only cylinders that are DOT certified refillable. It is a violation of the law to do otherwise.
3. Do not bump or drop refrigerant cylinders.
4. Mark the type of refrigerant that is in each cylinder. Use the cylinder only for that type of refrigerant.
5. Keep cylinders secured in an upright position.
6. Never apply heat (especially a welding torch) to a refrigerant cylinder. The cylinder could become overheated at the point where the flame touches the cylinder, weakening the cylinder wall and allowing an explosion.
7. Use only cylinders that are clean and have been evacuated.
8. Contaminated refrigerant must be cleaned to ARI 700-88 specifications by recycling or reclaiming. If this is not possible, it must be properly destroyed.

30

Health Hazards

The following information is a description of the safety considerations associated with fluorocarbon refrigerants. Before filling refrigerant cylinders or handling any fluorocarbon product, obtain and read thoroughly a copy of the product data sheets for the specific product.

CFC products usually are low in oral and inhalation toxicity, but possess properties that can pose significant health hazards under certain circumstances. Exposure to fluorocarbons above the recommended exposure levels (threshold limit values, or TLVs) can result in drowsiness and loss of concentration. Experiments with laboratory animals have demonstrated that cardiac arrhythmias can be induced by fluorocarbon levels of 20,000 to 100,000 parts per million. Fatal cardiac arrhythmias in humans accidentally exposed to high levels also has been reported.

It has been demonstrated in laboratory animals that fluorocarbons can sensitize the heart muscle to epinephrine (adrenaline). This may lower the threshold at which adverse affects (cardiac arrhythmia) occur when accidental overexposure occurs during strenuous physical activity. This factor also dictates that epinephrine not be administered as a medical treatment for any overexposure to these products.

CFCs are heavier than air and can displace the air from enclosed or semienclosed areas, resulting in a suffocation hazard. Do not fill refrigerant reclamation shipping containers in low or enclosed areas without proper ventilation.

Skin exposure to CFC products can result in irritation, primarily due to the fluorocarbon's defatting action. Skin or eye contact can also result in frostbite, primarily from liquefied or compressed gases. Use personal protection equipment, such as side shield glasses, gloves, safety shoes, and a hard hat when filling and handling reclamation containers.

Do not expose CFC products to sparks, open flame, lit cigarettes, or hot surfaces (250°F/121.1°C). Many fluorocarbon materials will decompose if exposed to temperatures above this level or open flames in the presence of air. The thermal decomposition products include hydrochloric acid and hydrofluoric acid vapors, which are both irritating and toxic. In addition, carbonyl halides (i.e., phosgene) can theoretically be produced during decomposition. These materials are toxic. Do not apply open flame to or heat reclamation containers above 125°F (52°C).

31

The following suggestions will help in the safe handling of refrigerant.

1. Always read the product label and the product safety data sheet.
2. Always use adequate ventilation.
3. Never expose these products to flames, sparks, or hot surfaces.
4. **Physicians**: Do not use epinephrine to treat overexposure.

Summary

◆ The Egyptians used water-filled clay containers to keep foods cool. As water slowly seeped through the porous walls and dried on the outside, this drying action, known as evaporation, caused the jar and its contents to cool. The same principle is the basis of modern mechanical refrigeration.

◆ More than a hundred years ago an English scientist succeeded in changing ammonia gas to a liquid by applying pressure and lowering the temperature. When the pressure was released, the ammonia boiled off rapidly and changed back to a gas. When this happened the scientist found that heat was removed from the surroundings. This discovery was important in the development of the modern mechanical refrigeration unit.

♦ Modern types of refrigerants are generally divided into three specific categories: chlorofluorocarbons, hydrochlorofluorocarbons, and hydrofluorocarbons.

♦ Chlorofluorocarbons are some of the most useful chemical compounds ever developed. They are nonflammable, noncorrosive, not very toxic, and are compatible with many types of materials.

♦ Chlorofluorocarbon refrigerants are reported to be causing damage to the ozone layer. They consist of chlorine, fluorine, and carbon. They have no hydrogen in their composition and are, therefore, very stable chemically. They have a very high ozone depletion potential.

♦ The leading candidate for the replacement of chlorofluorocarbon refrigerants are hydrofluorocarbon and hydrochlorofluorocarbon refrigerants.

32

♦ The hydrofluorocarbon refrigerants do not contain chlorine and have a zero ODP.

♦ The hydrochlorofluorocarbon refrigerants have hydrogen added to the chemical makeup of CFC refrigerants. This allows the dissipation of virtually all of the chlorine in the lower part of the atmosphere, before it has a chance to reach the ozone layer. Dissipation at lower altitudes gives these refrigerants a much lower ozone depletion potential, in the range of 2% to 10% of that of the CFC refrigerants.

♦ HFCs and HCFCs have a low toxic rate, stability, no or low flammability characteristics, and produce no smog and require minimal to moderate equipment changes when compared to not-in-kind refrigerants.

♦ An important consideration in the development of HCFC and HFC refrigerants as alternative refrigerants is the testing of oils for solubility, stability, lubricity with the refrigerant, and compatibility with materials of construction.

♦ HFC-134a is the alternative for CFC-12 in automotive air conditioning and many stationary refrigeration and air conditioning applications.

♦ Laboratory test results indicate that HFC-134a/PAG solutions have acceptable chemical stability. Other tests confirm that the HFC-134a molecule is at least as stable chemically as CFC-12.

♦ Ternary refrigerant blends are considered strong alternative candidates for CFC-12 in certain applications, including the automotive air conditioning aftermarket, refrigerators and freezers, and other stationary applications. Two blends, KCD-9430 and KCD-9433, were used for the oil testing.

◆ Lubricity tests show that ternary blends/alkylbenzene mixes do not perform well without oil additives. However, mixtures of blends and alkylbenzene lubricants respond well to the types of extreme pressure additives that have been used in refrigeration and air conditioning systems for many years.

◆ Both ternary blends and HFCs can be used with oils that are commercially available. However, some system changes or lubricant modifications may be required to optimize performance.

◆ Refrigerant cylinders are manufactured in a variety of types and sizes and are disposable or returnable.

◆ The type of refrigerant in a cylinder usually is determined by the color of the cylinder. The recovery cylinder has a yellow top and a gray body.

◆ All refrigerant cylinders are manufactured to meet specifications established by the U.S. Department of Transportation.

◆ Disposable cylinders (DOT 39s) must be tested and meet the requirements for the highest pressure refrigerant encountered, which is R-502. This calls for a service pressure of 260 psi test pressure. The test pressure for disposable cylinders is 350 psi, and they must not rupture below 650 psi.

33

◆ Disposable cylinders are manufactured from steel products and are subject to rusting. If the cylinder walls become rusty they may not be able to hold the pressures required of them. Thus, the use of disposable cylinders for compressed air tanks is hazardous.

◆ It is illegal to transport a DOT 39 cylinder that has been refilled. This illegal act has a maximum penalty of $25,000 and five years in prison.

◆ Recovery cylinders must comply with DOT specifications. These special cylinders are used for transporting contaminated refrigerant or refrigerant that has been recovered from a system for removal from the job site.

◆ Recovery cylinders are painted yellow on top and 12 inches down the side. The bottom of the cylinder is painted gray.

◆ Every cylinder has "MAX GROSS WT._____LB" stenciled on the shoulder. Never exceed the maximum weight.

◆ The DOT requires that every cylinder be manufactured with a safety relief valve.

◆ Do not fill a cylinder that is out of date. The test date appears on the tank. If it is found that it has been more than five years since the cylinder was tested, or retested, do not refill it. Return it to the owner for retesting.

♦ Never fill a cylinder more than 80% of its rated weight capacity at 70°F.

♦ When a refrigerant cylinder is to be subjected to temperatures in excess of 130°F, the cylinder should be filled only to 60% capacity because the refrigerant expands to fill the cylinder with liquid.

♦ Key Terms ♦

AEL

cardiac arrhythmia

CFC (chlorofluorocarbon)

DOT (U.S. Department of Transportation)

DOT 39

EPA (Environmental Protection Agency)

GWP (global warming potential)

HCFC (hydrochlorofluorocarbon)

HFC (hydrofluorocarbon)

Montreal Protocol

ODP (ozone depletion potential)

ozone layer

PAFTT (Program for Alternative Fluorocarbon Toxicity Testing)

PAG (polyalkylene glycol)

SSU (Saybold Seconds Universal)

TLV (threshold limit value)

VOC (volatile organic compound)

34

Review Questions

Essay

1. What is the principle used as a basis of modern mechanical refrigeration?

2. What was the first chemical refrigerant?

3. What is one of the most useful chemical compounds ever developed?

4. Which type of refrigerant is not very toxic, nonflammable, noncorrosive, and is compatible with many types of materials?

5. Which type of refrigerant is reported to be causing the most damage to the ozone layer?

6. Why are CFCs very stable chemically?

7. Why are CFCs being banned?

8. What are the leading candidates for the replacement of CFC refrigerants?

9. Which type of replacement refrigerants have a zero ozone depletion potential?

10. What component added to CFC refrigerants lowers their ozone depletion potential?

Fill-in-the-Blank

11. CFC refrigerants having _____ added to their chemical makeup have an ozone depletion potential of from 2% to 10% of the CFC types.

12. _____ and _____ have a low toxic rate, have stability, have either no or low flammability characteristics, produce no _____, and require minimal to moderate changes to the equipment.

13. An important consideration in the development of replacement refrigerants is the _____.

14. _____ refrigerant is the alternative refrigerant for _____ in automotive air conditioning systems and many stationary refrigeration and air conditioning applications.

15. _____ refrigerant blends are considered to be strong alternate candidates for some _____ applications.

True-False

16. Mixtures of blends and alkylbenzene lubricants do not respond well to the type of extreme pressure additives used in refrigeration and air conditioning systems for years.

17. The types of refrigerant cylinders are disposable and returnable.

18. A refrigerant recovery cylinder has a yellow top and the body is painted the color designating the new refrigerant type.

19. Refrigerant cylinders are manufactured to meet SAE specifications.

20. The test pressure for disposable cylinders is 650 psi.

21. Disposable cylinders should never be used for compressed air tanks.

35

Multiple Choice

22. A DOT 39 cylinder cannot be
    a. refilled.
    b. sold to another person.
    c. disposed of improperly.
    d. transported empty.

23. Every cylinder has stenciled on its shoulder
    a. "MAX. GROSS WT. _____ LB."
    b. "MAX. GROSS WT. _____ KG."
    c. "MAX. GROSS WT. - - - - - - EXCEEDED."
    d. the type of refrigerant inside.

24. Recovery cylinders are painted yellow on top and
    a. the same color as new refrigerant on the bottom.
    b. silver on the bottom.
    c. gray on the bottom.
    d. white on the bottom.

25. The DOT requires that every cylinder be manufactured with a
    a. reusable valve.
    b. pressure relief valve.
    c. strong handle.
    d. refillable valve.

26. A cylinder that is found to be out of date should be
    a. discarded.
    b. refilled with a note to have it retested.
    c. returned to the owner for retesting.
    d. used as usual with no special action.

27. At 70°F, a refrigerant cylinder should be filled to only
    a. 80% of its weight rated capacity.
    b. 70% of its weight rated capacity.
    c. 60% of its weight rated capacity.
    d. 50% of its weight rated capacity.

28. At 130°F or higher, a refrigerant cylinder should be filled to only
    a. 80% of its weight rated capacity.
    b. 70% of its weight rated capacity.
    c. 60% of its weight rated capacity.
    d. 50% of its weight rated capacity.

29. Before filling an empty refrigerant cylinder always
    a. evacuate the cylinder.
    b. purge the cylinder.
    c. pressure test it.
    d. check for the proper fill valve.

36

30. CFC products can safely be exposed to
    a. sparks.
    b. open flame.
    c. lighted cigarettes.
    d. air.

37

Refrigerant Regulations

Objectives

After completion of this chapter you should:

◆ *be more familiar with the current refrigerant regulations.*

◆ *be more familiar with the Montreal Protocol.*

◆ *know more about which chemicals are covered by the Clean Air Act Amendments of 1990.*

◆ *be more familiar with the 1990 update of the Montreal Protocol.*

◆ *know more about the Federal Taxes placed on CFC refrigerants.*

◆ *know the proposed refrigerant phaseout dates for CFC refrigerants.*

◆ *know the requirements set forth in Section 608 of Title VI of the Clean Air Act.*

◆ *know the technician's certification requirements set forth in Section 608 of the Clean Air Act.*

Recognizing the problems caused by depletion of the ozone, the industry started looking for safe substitutes for CFCs. Progress was being made when, in the early 1980s, the threat of more government regulation began to fade. At this time the search for refrigerant substitutes virtually stopped. The world use of CFCs continued to grow.

In 1985, scientists discovered a loss of ozone as large as North America over part of the southern hemisphere. Later data has revealed losses that exceed 50% in the total column and greater than 95% at an altitude of 9 miles to 12 miles above the earth. This newly discovered hole in the ozone gave new importance to international efforts to protect the ozone layer.

Early Controls on CFCs

The increased use of CFCs on a worldwide basis caused the EPA (Environmental Protection Agency) to consider some form of regulation on their use in the early 1970s. The theory of ozone depletion was new at that time, and few took it seriously.

The first restrictions placed on CFCs by the EPA and the FDA (Food and Drug Administration) were enacted in 1978. Restriction was placed on the use of CFCs for aerosol propellant applications, such as spray deodorants, spray paint, household cleaners, etc., when it was determined that approximately 50% of the CFCs produced were used for this purpose. There were only a few exceptions to this ruling, mostly in the medical field. This step also caused a decrease in the use of R-11 and R-12 for this purpose. However, only a few other countries joined the United States in this endeavor.

The EPA considered further restrictions on the production of CFCs in the 1980s. The EPA made two proposals for curbing the use of CFCs for other than aerosol use: (1) Mandatory regulations, and (2) economic incentive from storing and recycling. However, by this time the United States scientific community was losing interest in the ozone depletion problem. Because of this lack of interest, the EPA did not pursue these regulations at that time. The next step was an attempt to develop an international agreement to solve the problem.

Attention again focused on the two proposed regulations when global warming, or the "greenhouse effect," and new evidence that CFCs are in fact depleting the ozone layer captured the attention of the general public in 1986. It had been determined that the CFCs were causing a "hole" in the ozone layer over Antarctica. This caused major concern throughout the world. This concern led to what is now known as the Montreal Protocol, and later an update to this agreement.

39

Montreal Protocol

In September 1987, the Montreal Protocol on Substances that Deplete the Ozone Layer was negotiated and signed by more than two dozen nations. The Protocol was placed into force in January 1989. At the last count, 68 nations were parties to the Protocol.

Unfortunately, a short time after the Protocol was signed, scientists observed and measured losses of ozone on a worldwide scale. They also discovered that the loss was not limited to the remote, uninhabited portions of Antarctica. Losses over large areas of this country were recorded, bringing the problem closer to home.

These new measurements of the loss of the ozone layer were much greater than the computer models had predicted. Questions arose about the adequacy of control measures set forth in the original Montreal Protocol and the EPA regulations.

There are four major areas that need further attention by national legislatures and by the signing parties of the Protocol. These areas are accelerating the CFC and methyl chloroform phaseout schedules; controlling and ultimately eliminating the production and use of hydrochlorofluorocarbons (HCFCs); eliminating the emissions of ozone-destroying compounds; and implementing effective trade sanctions. Each of these areas is covered by the Clean Air Act Amendments of 1990.

These reassessments are to be scheduled regularly. The first, which took place in 1990, considered the following problem areas: effects of research; atmospheric sciences; and coverage and stringency of the protocol.

40

The United States Environmental Protection Agency enacted the provisions of the agreement to restrict the use of certain chemicals on August 1, 1988. This law placed restrictions on the chemicals that were determined to cause damage to the ozone layer.

Title VI of the new Clean Air Act, "Stratospheric Ozone Protection," represents one more link in the worldwide effort to safeguard the critical part of the atmosphere that protects the earth from harmful ultraviolet rays from the sun.

Chemicals Covered by the Montreal Protocol

The chemicals were divided into two separate groups: Group 1 consists of the fully halogenated chlorofluorocarbons; Group 2 consists of the halons. The following is a list of chemicals that were controlled by the original Montreal Protocol.

> Group 1 Fully Halogenated Chlorofluorocarbons
> > CFC-11
> > CFC-12
> > CFC-113
> > CFC-114
> > CFC-115
> Group 2 Halons (fire extinguisher agents)
> > Halon-1211
> > Halon-1301
> > Halon-2402

The original Montreal Protocol called for the manufacturing of CFC-11, CFC-12, CFC-113, CFC-114, and CFC-115 to be frozen at their 1986 consumption quantities. Restrictions were to begin approximately July 1, 1989, or within 90 days after the law was placed into force. Additional restrictions called for a 20% reduction from the 1986 consumption level by July 1, 1993, with further reductions of 50% by July 1, 1998.

The manufacture of halons 1211, 1301, and 2402 was to be frozen at 1986 consumption levels by 1992, or three years after the law was placed into force.

Each substance in both groups was assigned a number, or ozone depletion weight. The ozone depletion weight reflects the chemical's ability to destroy the stratospheric ozone layer. The ozone depletion weight is the same as the ozone depletion potential discussed earlier.

Update to the Montreal Protocol (1990)

On June 29, 1990, the first scheduled meeting of the signatories of the Montreal Protocol met in London, for the purpose of making the original protocol more stringent. Fifty-six countries, including the United States, attended the meeting, at which the schedule for phasing out CFCs was moved to an earlier date. Participants called for the phaseout of HCFCs in addition to the original phasing out of CFCs.

The revision of the protocol called for a decrease in production and consumption of CFCs from the 1986 usage level to meet the following schedule:

1/1/1993	20% reduction
1/1/1995	50% reduction
1/1/1997	85% reduction
1/1/2000	100% reduction (total phaseout)

At this same meeting it was decided that halons should be reduced by 50% by the year 1995 and completely phased out by the year 2000.

The call for phasing out HCFCs was to have them eliminated by January 1, 2040, and if possible by July 1, 2020. Some signatories wanted HCFCs phased out by the year 2000. Representatives of other countries resisted the phaseout of HCFCs by the year 2000 because the phasing out of CFCs would make HCFC refrigerants essential. The United States was among those that opposed rapid phasing out of HCFCs.

Federal Taxes on CFCs

The federal budget for 1990 contained provisions for placing an excise tax on CFCs and halons. This tax was effective January 1, 1990, and is to

41

be in effect until the complete phaseout of the listed CFCs. The CFCs to be taxed were CFC-11, CFC-12, CFC-113, CFC-114, and CFC-115. The tax is to be applied to each pound of these CFCs after the effective date and is to be charged to each pound of CFC that is sold or used by a manufacturer or an importer. Recycled and reclaimed refrigerants are to be exempt from these taxes.

Tax rates applied to these refrigerants are:

1990	$1.37	1993	$2.65
1991	$1.37	1994	$2.65
1992	$1.67	1995	$3.10

After 1995, the tax base amount for each pound of post-1989 and post-1990 CFCs will increase at a rate of $0.45 per pound per year.

The tax is to be paid with a Federal Tax Deposit Coupon, Form 8109, and deposited with any authorized depository or Federal Reserve Bank. Also, a tax return must be filed on the Quarterly Federal Excise Tax Return Form 720. All are to be attached to the Environmental Tax Form 6627.

The applicable tax is determined by multiplying the proper annual tax rate by the Ozone Depletion Weight (ODW) that has been assigned to each type of refrigerant, as follows:

Refrigerant	ODW
CFC-11	1.0
CFC-12	1.0
CFC-113	0.8
CFC-114	1.0
CFC-115	0.6

1990 Amendment to the U.S. Clean Air Act

On November 15, 1990, President Bush signed into law the Clean Air Act Amendment of 1990. In Title VI ("Stratospheric Ozone Protection") of the amendment, the United States Congress established a comprehensive method for phasing out ozone-depleting substances. The EPA is to develop regulations to implement these amendments to the Clean Air Act.

Title VI divides the ozone-depleting chemicals into two classes. Class I substances are the CFCs, halons, methyl chloroform, and carbon tetrachloride. Class II substances are the hydrochlorofluorocarbons.

Section 608 of Title VI covers the National Recycling and Emission Reduction Program. Title VII is the regulatory portion of the amendment.

The amended Clean Air Act does not require individual states to develop regulations to protect the stratospheric ozone layer. The federal regulations will override any such state or local regulation unless the state proves to the EPA that its regulations are at least as stringent as the federal regulations and that the state has the resources to enforce the regulations and is in fact enforcing the regulations.

Section 608 of the Clean Air Act requires that the United States Environmental Protection Agency (EPA) develop specific regulations prohibiting emissions of ozone-depleting chemicals during their use and disposal to the lowest achievable level (LAL) and to maximize recycling of the chemicals. The act also prohibits the release of refrigerants into the atmosphere during the servicing, maintenance, and disposal of refrigeration and air conditioning equipment starting July 1, 1992.

The EPA has developed regulations requiring the recycling of ozone-depleting chemicals (both CFCs and HCFCs) during the servicing, repair, or disposal of refrigeration and air conditioning equipment. These regulations also require that refrigerant in appliances, machines and other goods be recovered from these items prior to their disposal.

Technician Certification

43

Technicians who service, install, and maintain air conditioning and refrigeration equipment must be certified by the EPA. The recovery and recycling equipment used during service operations must also be certified by the EPA. The technician certification is divided into four (4) types, as follows: Type I (Small Appliances), Type II (High Pressure), Type III (Low Pressure), and Type IV (Universal). In the certification exams, there are twenty-five questions that are common to all four types of certification and are included on all certification exams. These questions focus on the environmental impact of CFCs and HCFCs, the regulations and changing the outlook of the industry, the filling and handling of refrigerant cylinders, and the exposure levels allowed in equipment rooms. Type I, Type II, and Type III certification exams have another twenty-five questions specific for each sector covered by the particular certification. Type IV certification exams have twenty-five questions from each of the Types I, II, and III, plus twenty-five questions common to all types of certification.

Technicians have until November 1994 to become certified. Then they may work only on the types of systems for which they are certified. When a technician is certified, there will be no recertification required. However, the EPA may require that technicians demonstrate proper techniques for recovery-recycling unit operation. If proper procedures are not followed, the technician may loose certification.

Type I (Small Appliances) Certification

The exam includes questions concerning recovery devices that are unique to the small appliance sector of the industry. Also included is system-dependent recovery techniques. This test may be taken either on-site or by mail. The required passing grade is 84% when taken by mail and 70% when taken on site.

Type II (High Pressure) Certification

The questions for this certification deal with vacuum levels required, proper use of recovery equipment for removing both liquid and vapor refrigerant from the system, the purpose of system receivers, and the use of refrigerant monitors in equipment rooms. This is a closed-book, proctored test. The required passing grade is 70%.

Type III (Low Pressure) Certification

44

This certification covers any type of equipment that contains a low-pressure refrigerant, such as CFC-11. The questions for this type of certification include evaporator leak testing methods and proper procedures for deep evacuating the system. This is a closed-book, proctored test. The required passing grade is 70%.

Type IV (Universal) Certification

The exam for this type of certification has twenty-five questions from each of the Types I, II, and III in addition to the twenty-five questions that are common to all four types of certification. The questions include shipping procedures, disposal and proper refrigerant cylinder handling. This is a closed-book, proctored test. The required passing grade is 70%.

Contractor Self-Certification

This is another type of certification that is required by the EPA. Those wishing this type of certification can obtain the proper form from the EPA. The form is then filled out and mailed to the appropriate EPA regional office.

Those who service air conditioning and refrigeration systems, and those who dispose of appliances, with the exception of small appliances and room air conditioners must self-certify that they either own or have leased certified recovery-recycling units. The deadline for self-certification is August 6, 1993.

Those persons, who recover refrigerant from small appliances and room air conditioners before disposing of them, must also self-certify that the recovery-recycling units are EPA approved.

Self-certifications are not transferable to a new owner. When a business changes ownership, the new owner has 30 days to self-certify the equipment used.

Evacuation Standards

It is required that systems containing a refrigerant charge of 200 pounds or more of a high-pressure refrigerant, such as HCFC-22, must be evacuated to 10 in. Hg during the recovery process and before the system is opened to the atmosphere for repairs, with the exceptions described in the next section. This requirement is because of the difficulty in reaching lower vacuums with these types of refrigerants.

Systems containing a charge of more than 200 pounds of CFC-12 or CFC-502 must be evacuated to 15 in. Hg during the recovery process and before the system is opened to the atmosphere for repairs, with the exceptions described in the next section.

Systems containing a charge of less than 200 pounds of CFC-12 or CFC-502 must be evacuated to 10 in. Hg during the recovery process, with the exceptions described in the next section.

Low-pressure equipment must be evacuated to 29 in. Hg during the recovery process and before the system is opened to the atmosphere.

Exceptions

There are two exceptions to the evacuation standards previously described.

1. Nonmajor repairs that, after they are completed, do not involve evacuation of the refrigerant to the atmosphere.
2. Leaks that will not allow the required evacuation to be reached.

It was decided that both of these instances would actually cause greater emissions when the required evacuation level was attempted than when less, or none at all, was used.

45

Nonmajor Repairs

This type of repair involves making a small opening for a short period of time, only a few minutes. A nonmajor repair allows only a small amount of refrigerant to escape and the amount of air and moisture that can enter the system is minimal. Also included in this category is the replacement of components such as safety and pressure switches and filter-driers.

When making repairs of this nature, the system can be evacuated to 0 psig for high-pressure equipment and pressurization to 0 psig for low-pressure systems. This involves performing service operations that do not require the evacuation of refrigerant to the atmosphere before being placed back in operation.

Major Repairs

This type of repair involves making large openings in the system. This includes such procedures as compressor replacement, condenser removal, evaporator removal, or an auxiliary heat exchanger removal. This category can be categorized by the need to evacuate the system after repairs and before charging the system for operation. A major repair is involved when refrigerant left in the system is purged through the vacuum pump to the atmosphere during the evacuation process.

46

Leak Repair Requirements

Systems that contain a refrigerant charge of 50 pounds or more must have any leak repaired within 30 days. Systems containing less than 50 pounds are not covered by this requirement. It is the responsibility of the equipment owner to have the repair completed if the annual leak rate exceeds 15% or 35% of the total charge as described in the next section. The technician should inform the owner of the leak and of the requirements by the EPA. The owner cannot intentionally ignore any information that reveals that a leak exists. This is to prevent topping-off the charge in a leaking system.

Flexibility

There is a certain amount of flexibility in the requirements concerning leak repairs. These are as follows:

◆ Annual Leak Rate of 15% or Higher. Air conditioning systems such as those used in commercial buildings and hotels must have the leaks repaired.

◆ Annual Leak Rate of 35% or Higher. This area covers equipment that is used in an industrial process, commercial refrigeration, pharmaceutical systems, petrochemical systems, chemical systems, industrial ice machines and ice rinks must have all leaks repaired.

The technician can estimate the size of the leak by checking past invoices for service procedures requiring refrigerant to be charged into the system. There are probably other means of estimating the leak size if the unit is maintained by a service company or a maintenance department.

There is one exception to the leak repair requirements. If the owner decides to replace or retrofit a leaking system but must wait a couple of months before making the necessary changes, the owner must develop a detailed plan within 30 days of being informed showing his intentions to replace or retrofit the equipment. This plan must be dated and kept on-site and subject to EPA inspection. The repair or retrofit must be completed within one year from the date at which the plan was initiated.

47

Refrigerant Phaseout Dates

The following list and Figure 3-1 show projected production phaseout dates and quantities for CFCs and HCFCs.

CFC production is to be reduced from 1986 production levels by the year and percentage shown.

1991	15%	1996	60%
1992	20%	1997	85%
1993	25%	1998	85%
1994	35%	1999	85%
1995	50%	2000	100%

HCFC phaseout schedule:

2015	Production and use is to be frozen and limited
2020	No new refrigeration or air conditioning uses
2030	100% phased out

Starting January 1, 1992, use of recovering and recycling equipment by anyone servicing automotive air conditioning systems for pay became mandatory.

Starting July 1, 1992, there will be no intentional venting of CFCs or HCFCs from any type of refrigeration or air conditioning system using these refrigerants. The refrigerants must be recovered, recycled, or reclaimed according to Section 608 of the Title VI Amendment of the Clean Air Act.

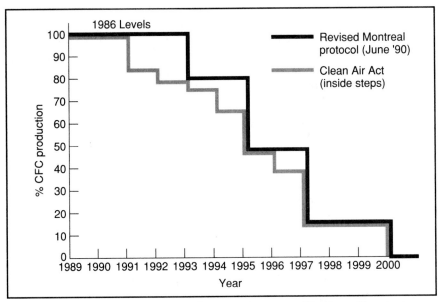

Figure 3-1 CFC regulations on production. *Courtesy of Du Pont Chemicals.*

Compliance dates

The following are the compliance dates and what must be done to meet EPA guidelines:

June 14, 1993. When a leak of 35% or more per year is found on a system containing a refrigerant charge of 50 pounds or more, such as commercial refrigeration or an industrial processing unit, the system owner must be notified. The owner must then repair the unit within 30 days.

When a leak of 15% per year is found on any type of system containing a refrigerant charge of 50 pounds or more, the owner is responsible for having it repaired within 30 days.

If the owner develops a replacement or retrofit plan within 30 days of notification, there is a one year exemption from making the repairs. The plan must be kept on-site and available for EPA inspection.

July 13, 1993. After this date to legally open a refrigeration system, the service company must have at least one self-contained, EPA certified recovery unit at the business location.

If a major repair is to be performed on any high-pressure system with a refrigerant charge of 50 pounds or more, it must be pumped down to a vacuum of 10 in. Hg before the system is opened. If a nonmajor repair is performed or if evacuation would substantially contaminate the recovered refrigerant, the system can be evacuated to 0 psig. The evacuation to 0 psig is also applicable if the system is not to be evacuated after the repair and before recharging the unit.

Before opening a low-pressure system, the pressure inside the system must be increased to 0 psig. This is to prevent pulling moisture and air into the system and contaminating the system and refrigerant.

When evacuating small appliances, such as refrigerators, freezers, room units, packaged terminal air conditioning units, packaged thermal heat-pump systems, dehumidifiers, vending machines, drinking water coolers, or units containing a refrigerant charge of 5 pounds or less, the level of evacuation depends on whether or not the system compressor is operational.

If the system compressor is operational, both the self-contained and the system-dependent recovery units must produce a 90% evacuation level. When the system compressor is not operational, the recovery procedure must remove 80% of the charge.

When disposing of a system containing refrigerant of any amount, an EPA approved recovery-recycling unit must be used to evacuate the system to the level discussed.

Any refrigerant removed from a system, or a part of a system to be serviced, must be removed with an EPA approved recovery-recycling unit.

When disposing of a small appliance the last person to handle the appliance must recover any refrigerant or verify in writing that it has been recovered.

August 12, 1993. After this date it is illegal to open any refrigerant system for repair, maintenance, or disposal unless the technician has certified to the EPA the ownership of the EPA certified recovery-recycling equipment. The certification form is available from EPA.

It is also illegal to sell any refrigerant to a new owner unless it has been reclaimed and certified to meet ARI Standard 700 purity. The used refrigerant can be used in another system with the same owner without prohibition.

Reclaimers must certify that the refrigerant has been cleaned to meet ARI 700 specifications and that not more than 1.5% of it will be vented during the reclamation process.

Reclaimers must report to the EPA within 45 days from the end of the calendar year regarding the quantity of refrigerants received and

49

reclaimed. They must also furnish the names of the persons who supplied the refrigerant to them.

November 15, 1993. After this date all recovery-recycling units must be certified by the EPA to meet the approved evacuation levels.

After this date it is also illegal to sell anything but small appliances that are not equipped with a servicing aperture and a process tube so that the refrigerant can be recovered from the system.

The manufacturers of recovery-recycling units must have their units certified by an approved testing organization. The equipment must be able to meet the evacuation levels stated.

The recovery and recycling of refrigerants represents a critical step in protecting the stratospheric ozone layer. As an encouragement for owners and users of air conditioning and refrigeration equipment to start recovering and/or recycling CFCs as soon as possible, the EPA will grandfather recovery and recycling equipment. However, these provisions require that the equipment certified before January 1, 1994, must pump a vacuum of 4 in. Hg for high-pressure systems and a vacuum of 25 in. Hg for low-pressure systems.

50

November 14, 1994. After this date, a technician cannot legally open, or dispose of refrigeration equipment unless proper technician certification has been attained. Only training-testing organizations that are EPA approved can train and certify technicians after this date. It is also illegal to sell refrigerant to anyone except EPA certified technicians.

Summary

- The first restrictions placed on CFCs by EPA and FDA were enacted in 1978.

- Restriction was placed on the use of CFCs for aerosol propellants, when it was determined that approximately 50% of the CFCs produced were used for this purpose.

- In the 1980s, the EPA made two proposals concerning the use of CFCs for other than aerosol use. These proposals were economic incentives for storing CFCs and mandatory regulations. The EPA did not pursue these regulations at that time.

- In September, 1987, the Montreal Protocol on Substances that Deplete the Ozone Layer was negotiated and signed by more than two dozen nations. The protocol went into effect in January, 1989.

- Scientists discovered that the ozone layer loss was not limited to the remote, uninhabited portions of Antarctica.

◆ New measurements of the loss of the ozone layer were much greater than the computer models had predicted. Therefore, questions were raised about the adequacy of control measures set forth in the original Montreal Protocol and the EPA regulations.

◆ Four areas need further attention by national regulators and the signing parties of the protocol. They are accelerating the CFC and methyl chloroform phaseout schedules; controlling and ultimately eliminating the production and use of HCFCs; eliminating the emissions of ozone destroying compounds; and implementing effective trade sanctions. Each of these is covered in the Clean Air Act Amendment of 1990.

◆ Title VI of the new Clean Air Act represents one main link in the worldwide effort to safeguard the critical part of the atmosphere that protects the earth from harmful ultraviolet rays from the sun.

◆ The chemicals covered by the Montreal Protocol were divided into two groups: Group I consists of fully halogenated chlorofluorocarbons; Group II consists of the halons.

◆ The original Montreal Protocol called for the manufacturing of CFC-11, CFC-12, CFC-113, CFC-114, and CFC-115 to be frozen at their 1986 consumption quantities. These restrictions were to begin approximately July 1, 1989, or at 90 days after the law went into effect. Additional restrictions called for a 20% reduction from the 1986 consumption level by July 1, 1993, and reductions of 50% by July 1, 1998.

51

◆ The manufacture of halons 1211, 1301, and 4202 also was to be frozen at 1986 consumption levels by 1992, or three years after the law went into effect.

◆ In 1990 the first scheduled meeting of the signatories of the Montreal Protocol met in London to make the original protocol more stringent.

◆ The federal budget for 1990 contained provisions for an excise tax on CFCs and halons. This tax was effective January 1, 1990, and is to remain in effect until the complete phaseout of listed CFCs. Recycled and reclaimed refrigerants are exempt.

◆ Section 608 of Title VI covers the National Recycling and Emission Reduction Program. Title VII is the regulatory portion of the amendment.

◆ The amended Clean Air Act does not require individual states to develop regulations to protect the stratospheric ozone layer. Federal regulations override state or local regulations unless the state proves to the EPA that its regulations are at least as stringent as the federal regulations and that the state is enforcing the regulations.

◆ Starting July 1, 1992, intentional venting of CFCs or HCFCs from any type of refrigeration or air conditioning system using these refrigerants will be prohibited. These refrigerants must be recovered, recycled, or reclaimed according to Section 608 of Title VI Amendment of the Clean Air Act.

♦ The four types of technician certification are Type I (Small Appliances), Type II (High Pressure), Type III (Low Pressure), and Type IV (Universal).

♦ There are twenty-five questions on the certification exams that are common to all four types of certification. There are also twenty-five questions specific for each sector for each of Type I, Type II, and Type III exams. There are seventy-five questions including twenty-five from each of Type I, Type II, all Type III on the Type IV exam plus the twenty-five questions common to all four types of certification.

♦ Key Terms ♦

CFC (chlorofluorocarbon refrigerant)

EPA (Environmental Protection Agency)

FDA (Food and Drug Administration)

greenhouse effect

HCFC (hydrochlorofluorocarbon refrigerant)

LAL (lowest achievable level)

Montreal Protocol

ODW (ozone depletion weight)

stratospheric ozone layer

52

Review Questions

Essay

1. When were the first restrictions placed on CFCs?

2. For what were the most CFCs used in the early years of regulation?

3. When was the Montreal Protocol negotiated?

4. When was the Montreal Protocol placed into effect?

5. What caused a renewed interest in the depletion of the ozone layer?

6. In what law was eliminating the emissions of ozone-destroying compounds addressed?

True-False

7. Title VI of the Clean Air Act is named "Stratospheric Ozone Protection."

8. Halons fall into the Class I substances of the Montreal Protocol.

9. The Montreal Protocol called for manufacturing restrictions on the manufacture of CFC refrigerants by July 1, 1990.

10. The manufacture of halons was to be frozen at their 1990 production levels.

Fill-in-the-Blank

11. The Montreal Protocol placed chlorofluorocarbons into Group __.

12. Chlorofluorocarbon refrigerants are frozen at their _____ consumption levels.

13. The first scheduled meeting of the parties of the Montreal Protocol was for the purpose of making the original protocol more _____.

14. The Federal Excise Tax on CFCs and halons is to be in effect until the _____ phaseout of the listed CFCs.

15. Recycled and reclaimed refrigerants are to be exempt from _____.

Multiple Choice

16. The tax base amount per pound of post-1989 and 1990 CFCs will increase by
    a. $0.45 per pound per year.
    b. $0.75 per pound per year.
    c. $1.45 per pound per year.
    d. $1.75 per pound per year.

53

17. The National Recycling and Emission Reduction Program is presented in
    a. Section 608 of Title VII.
    b. Section 609 of Title VII.
    c. Section 608 of Title VI.
    d. Section 609 of Title VI.

18. The enforcement of the Clean Air Act rests with the
    a. city EPA.
    b. county EPA.
    c. state EPA.
    d. federal EPA.

19. There can be no intentional venting of CFCs or HCFCs from any refrigeration or air conditioning system after
    a. January 1, 1992.
    b. July 1, 1992.
    c. January 1, 1993.
    d. July 1, 1993.

20. The certification required to service residential air conditioning systems is
    a. Type I.
    b. Type II.
    c. Type III.
    d. Type IV.
21. Systems holding a refrigerant charge of less than 200 pounds of CFC-12 or CFC-502 must be evacuated to
    a. 10 in. Hg.
    b. 15 in. Hg.
    c. 22 in. Hg.
    d. 29 in. Hg.
22. How long does a customer have to repair a leak in a system with a refrigerant charge of 50 pounds or more of refrigerant?
    a. requires immediate repair.
    b. 15 days.
    c. 20 days.
    d. 30 days.
23. Technicians must be certified by
    a. July 1, 1993.
    b. July 1, 1994.
    c. November 14, 1994.
    d. January 1, 1995.
24. Repairs that do not require evacuation of the refrigerant to the atmosphere are known as
    a. simple.
    b. common.
    c. major.
    d. nonmajor.

Chapter 4

Refrigerant Conservation

Objectives

After completion of this chapter you should:

♦ *be more familiar with refrigerant conservation.*

♦ *be more familiar with the EPA regulations regarding intentional venting of CFCs and HCFCs.*

♦ *know more about the service procedures and practices used in the conservation and containment of refrigerants.*

It has been estimated that the CFCs in use today and those manufactured in the future, even with the suggested accelerated phaseout, with proper conservation will last for many more years. However, for this to become a fact everyone must practice proper conservation techniques by using good service procedures as long as CFC systems are still in use.

Availability

With a phaseout schedule in place, the conservation of CFCs is very important. CFCs will be scarcer and more difficult to purchase. The suggested phaseout by the year 1995 rather than the year 2000 will make CFCs even more difficult to obtain. But proper conservation tech-

55

niques will allow new refrigerants to be available for many years. This will be possible due to proper recovery, recycling, and reclaim practices.

Economics

The limited availability of new refrigerants is only part of the reason to conserve. One that cannot be overlooked is that of economics. This will be more apparent when one considers the tax on each pound of new refrigerant. The tax does not apply to refrigerant that has been recovered, recycled, or reclaimed. Only new refrigerant is taxed. By the year 1993 the tax will be $3.10 per pound of refrigerant and $0.45 per pound each year thereafter. This likely is strong incentive to practice conservation. New refrigerant would be used only in situations when a loss of refrigerant cannot be avoided, such as line breaks and equipment replacements.

56

EPA Regulations

After July 1, 1992, the Clean Air Act prohibits the intentional venting of ozone-depleting chemicals, such as CFCs and HCFCs, from stationary air conditioning and refrigeration systems. The EPA has been charged with the responsibility to reduce the use and emissions of CFC and HCFC refrigerants to the lowest achievable level (LAL). Under these sanctions, starting July 1, 1992, all refrigerant leaks in a system must be located and repaired within 30 days or a written plan, dated and kept on-site, must be prepared showing the intention to replace or retrofit the equipment. The intentional venting of CFCs and HCFCs after July 1, 1992, carries a fine of up to $25,000 per day per offense. Each kilogram of refrigerant is counted as an offense. There is also an Awards Provision that allows anyone to report an offender. If the offender is convicted, the person who reported the violation receives up to $10,000. At these rates for violation, the price for recovery, recycling, or reclaim equipment can be paid for quickly.

Code of Ethics

A code of ethics, such as the one by Carrier Corp., is worth practicing, Figure 4-1. It establishes standards for the safe, efficient handling of refrigerants.

OBJECTIVE:

To Provide an Industry-Wide Code of Practice for the Servicing of Air Conditioning/Refrigeration Equipment and Systems for the Purpose of Conserving Refrigerant, Improving Personal Safety, and Preserving the Environment.

DO:

❏ Think CFC conservation and safety.

❏ Follow and use recommended procedures and equipment for handling refrigerants.

❏ Educate our customers by explaining the Code of Practice and by offering seminars and training programs.

❏ Replace and tighten all seal caps on all valves after servicing.

❏ Shut down system and make repairs when leaks exist.

❏ Use closed loop refrigerant transfer equipment when removing, charging, and storing refrigerants.

❏ Recover vapor and liquid refrigerant from charging hoses.

❏ Maintain refrigerant use log for all equipment. (Record on-site, as well as service report.)

❏ Leak test all charging hoses and refrigerant handling equipment.

❏ Install service isolation valves to limit refrigerant losses during servicing (+ purge).

❏ Eliminate unnecessary mechanical joints. Use welded or brazed joints.

❏ Establish proper leak testing routing.

❏ Follow published leak test procedure (pressure vessels and oil cooler).

❏ Use industry accepted tools/equipment for leak testing.

❏ Confirm overall leak tightness by using a standing vacuum test.

❏ After major service, evacuate and dehydrate to minimum 29.8 inches by using deep vacuum or triple evacuating method.

❏ Install purge counters.

❏ Install hour meters on all equipment.

❏ Install more efficient purges that reclaim exhaust vapor.

❏ Install external oil filters.

❏ Elevate oil temperature prior to service work.

❏ Run auxiliary oil pump weekly to flood oil seal on open-drive systems.

❏ Use only approved cylinders/drums/tanks for storing refrigerant.

❏ Install charging valve quick connects.

❏ Cool refrigerant drums to atmospheric pressure prior to opening.

57

continued on next page

Figure 4-1   Code of service practice. *Courtesy of Carrier Corporation.*

CODE OF SERVICE PRACTICE—*Continued*

DO:

❏ Install refrigerant sensors on/near all refrigerant systems.

❏ Recover all refrigerant for recycling/reclaiming.

❏ Use non CFC gas as tracer gas when leak testing (i.e., HCFC 22).

❏ Install alarm system to warn of excessive machine pressure during shutdown.

❏ Use purge compressor or portable evacuation device to recover refrigerant liquid/vapor from refrigerant drums/cylinders.

❏ Add refrigerant carefully to avoid overcharging.

❏ Calibrate controls with air, nitrogen, or control calibration sets.

❏ Inspect for abnormal vibration.

❏ Implement an effective water treatment control.

❏ During periods of chiller shutdown (low pressure units), utilize slight positive pressure devices or store refrigerant in a manner conducive to refrigerant conservation.

❏ Dispose of used refrigerant containers properly.

DON'T:

❏ Use refrigerant to clean tools, coils, machinery or to blow off pipes, etc.

❏ Use refrigerant as cleansing solvent (compressor clean ups).

❏ Open the refrigerant side of system unless "absolutely" necessary.

❏ Use CFC as tracer gas for leak testing.

❏ Operate equipment known to have leaks.

❏ Cool bearings and parts for fitting with refrigerants.

❏ Vent/blow off air (non condensables/refrigerants) to atmosphere.

❏ Blow off "empty" tanks, drums or containers.

❏ Blow off vapor still in chiller after liquid removal.

❏ Pressurize chiller with air if refrigerant is still in chiller.

❏ Mix refrigerants, solvents, oil, for disposal.

❏ Exceed manufacturers recommended pressure when leak testing.

❏ Throw away any refrigerant. (Reclaim, recycle, recover, reuse.)

❏ Operate chiller in "surge" conditions.

❏ Overfill refrigerant containers, tanks, drums, recovery units, receivers, etc.

❏ Refill disposable cylinders.

❏ Substitute alternative refrigerants into old systems without approval.

Printed on Recycled Paper. 808-742 0991

Carrier
BUILDING SYSTEMS AND SERVICES
Customer Satisfaction, Our First Priority

58

Figure 4-1   Code of service practice. *Courtesy of Carrier Corporation.*

Service Procedures and Practices

The conservation and containment of refrigerants have become much more a part of service procedures and practices since July 1, 1992. Service technicians will become more aware of waste prevention.

Reduce Waste

There must be less refrigerant wasted during the leak test and other service procedures, and purging off of a small amount of refrigerant to adjust the system charge must be stopped.

Only 5% of the total system charge can be vented purposely during service operations involving refrigerant. Using valves, gaskets, hoses, and other such wasteful devices will be a thing of the past. The use of refrigerants to blow out drain lines and condenser coils, and other such uses will not be permitted. Technicians will soon learn to use other ways of doing the service operations for which refrigerant has been used in the past. On sealed systems having no service valves, solder-on type line piercing valves can be used successfully to gain access to the refrigerant cycle. Bolt-on line piercing valves have a tendency to leak, causing future problems.

59

Identify and Repair Leaks

Keeping the refrigerant in the system is more important now than ever. It can do no harm to the atmosphere if it is not allowed to escape. More time will be used in finding and repairing leaks and in making certain that repair work is first class and that proper procedures are used in refrigerant conservation. Careful consideration must be given to gaskets, flare joints, soldered joints, and points at which refrigerant tubing may rub against something that will cause wear, allowing the refrigerant to escape. The seals on open compressors need to be kept properly lubricated when they are out of service for extended periods. Extra time and quality parts will become the norm rather than the exception.

The most common methods for detecting leaks in CFC and HCFC refrigerant systems are the visual check, bubble test, halide leak detector, electronic leak detector, and chemical dyes. Use the following pressurization procedure for all leak testing methods discussed.

Pressurize the System. When testing for leaks, make certain that the pressure in the system is sufficient to cause a leak during the test. If the system contains refrigerant, pressurize the system to the saturation pressure for the type of refrigerant in the system. When the leak or leaks are found, the refrigerant is removed from the system using the recovery unit. Do not vent this refrigerant to the atmosphere. Only 5% of the total system charge may be wasted for service operations.

When testing for leaks in a completely empty system, do not fill the system with refrigerant and then purge the refrigerant to make the repair. Pressurize the system with dry nitrogen to near operating pressure, and use the bubble test. If no leaks are found, introduce a small amount of HCFC-22 into the system and test with either the halide or the electronic leak detector. This nitrogen and refrigerant can then be recovered and placed in a separate cylinder (not mixed with other refrigerants), and saved for future leak testing. It is recommended that this nitrogen/HCFC-22 mixture be dried before using it in another system to prevent possible contamination. All suspected leaks should be tested and repaired at one time so that only one refrigerant recovery operation is needed.

When the leak, or leaks, are located, use the recovery unit to remove remaining refrigerant from the system. Repair the leaks and evacuate the system to 500 microns, or to the equipment manufacturer's specifications. Then recharge the system. The recovered refrigerant can be recycled and charged back into the system if it is not severely contaminated.

60

Visual Check. When a system is found to be short of refrigerant, a visual check will sometimes locate the problem without much trouble. The leak usually has refrigeration oil around it. Any time oil is on a joint, a leak should be suspected. When the leak is repaired, the oil should be removed so the next technician will not suspect a leak. Use the procedure noted earlier to complete the service operation.

Bubble Test. To use the bubble test, it is usually best to obtain a bottle of liquid plastic (known as soap bubbles). A rich mixture of liquid dish washing soap and water may also be used for this test; however, liquid plastic is preferred. Wet the area around the suspected leak and observe the joint for at least 15 seconds. If a leak is present bubbles will appear at the leak. The refrigerant can then be recovered from the system and the leak repaired. Use the procedure noted earlier to complete the service operation.

Halide Leak Detector. This type of leak detector draws a small quantity of refrigerant through a tube and causes it to pass over a heated element. The refrigerant then turns color: greenish-blue for a small leak and bright purple for a large leak. Place the end of the hose around all suspected places until the leak, or leaks, are found. Be careful when using a halide leak detector because the chimney around the element gets very hot. Also do not breathe the vapors from the flame. These vapors are very toxic. Use the procedure noted earlier for the remainder of the operation.

Electronic Leak Detector. These devices are sensitive and cannot be used successfully in areas having high concentrations of refrigerant. Electronic detectors draw the refrigerant through a tube and over a heated element. When the end of the tube is passed over a leak, a light glows, a beeping noise is heard, or both a light and a beeping sound are activated. Electronic leak detectors should be used according to the manufacturer's recommendations. Use the procedure noted for the remainder of the operation.

Chemical Dyes. Chemical dyes are introduced into the system and circulated with the refrigerant. When a leak is present, the dye escapes with the refrigerant and usually causes the area around the leak to turn red. Any red coloration indicates the presence of a leak. Be sure to check all the places where a leak might occur before removing the refrigerant. Some equipment manufacturers state that dyes should not be placed in their units. Be sure to check with the equipment manufacturer before using dye in the system. Use the procedure noted earlier for the remainder of the operation.

Containment

61

An estimated 75% of the refrigerant used today is for service operations, Figure 4-2. Therefore, keeping it inside the system (containment) is the first step in CFC conservation.

Alternatives eventually will replace the chlorine-based refrigerants. In the short term, it is important to recognize that refrigerants cannot dam-

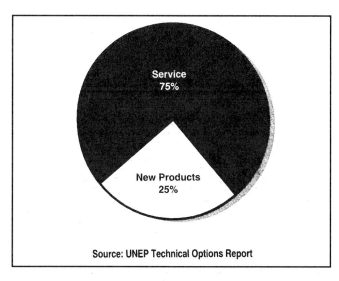

Figure 4-2   Estimated annual refrigerant usage. *Courtesy of Carrier Corporation.*

age the environment unless they escape into the atmosphere. By applying technologies to contain refrigerants today, we can dramatically slow refrigerant-related environmental problems without waiting for tomorrow's solutions. We also can ensure an adequate CFC and HCFC refrigerant supply to service existing equipment.

Containment of all refrigerants, whether regulated or not, is common sense from a business standpoint and will make even more sense as refrigerant prices rise.

Negative-Pressure Equipment

Negative-pressure equipment includes systems that operate with pressures at or below atmospheric pressure. These types of equipment pose special problems because they tend to draw in air and moisture while operating and then sometimes leak refrigerant out during the idle period. Special devices are used to prevent the escape of CFCs into the atmosphere during the idle time.

Modernization

62

Modernization is one answer for existing equipment. Most refrigerant losses from operating or idle equipment can be eliminated by the addition of aftermarket devices such as high-efficiency purges, vacuum preventing devices, and pressure relief valves.

High-Efficiency Purges. Most refrigerant losses from operating or idle equipment can be eliminated by the addition of a high-efficiency purge, Figure 4-3.

This type of high-efficiency retrofit purge can cut refrigerant losses by up to 99%.

Vacuum Preventing Control. These devices prevent the formation of a vacuum during chiller shutdown or standby, helping to eliminate not only CFC loss and air/moisture contamination, but also hard start-ups and acid damage to the machine, Figure 4-4.

Pressure Relief Valves. Rupture discs can be retrofitted with pressure relief valves to prevent the accidental total loss of CFC refrigerant charge, Figure 4-5, page 58.

Tools

New technologies and improved service practices will help to avoid refrigerant losses. Common sense, commitment to improving the envi-

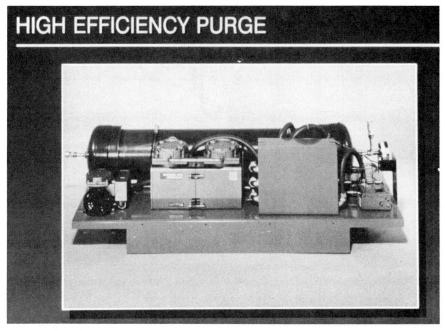

Figure 4-3   High-efficiency retrofit purge. *Courtesy of Carrier Corporation.*

63

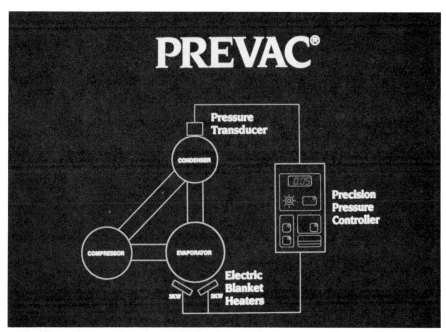

Figure 4-4   Vacuum-preventing control. *Courtesy of Carrier Corporation.*

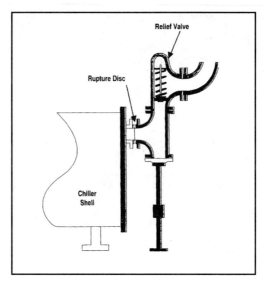

Figure 4-5   Pressure relief valve. *Courtesy of Carrier Corporation.*

64

ronment, and some simple new tools extend the useful lives of chillers and reduce operating costs by ensuring an adequate supply of refrigerants, regardless of the phaseout dates.

Among the tools and practices are refrigerant management systems, prevac systems, oil-monitoring service, purge units, and a code of ethics.

Refrigerant Management Systems (RMS)

A properly designed refrigerant management system recovers the refrigerant charge (including the vapor) of negative-pressure chillers, recycles it, then returns it to the chiller, Figure 4-6.

Prevac

These devices, when used as a service tool, can pressurize the refrigerant circuit in negative-pressure chillers so that the system can be checked for leaks without adding nitrogen or air, which then would have to be removed (see Figure 4-4).

Oil-Monitoring Service

These devices are used for periodic monitoring of the oil quality in the chiller to eliminate unnecessary oil changes which pose a risk of refrigerant loss to the atmosphere.

Figure 4-6   Refrigerant management system. *Courtesy of Carrier Corporation.*

Purge Unit

Purge units are a small part of chillers and other refrigeration systems, but they can be a major source of refrigerant emissions if they do not operate properly.

By removing noncondensable gases that could keep the condenser pressures extremely high, purge units improve refrigeration system efficiency.

Mechanical purge units consist of an inlet pressure control valve located on the suction line between the system condenser and the purge unit compressor. The compressor discharges gas at a higher pressure to an optional oil trap and then into the condenser, Figure 4-7.

Gases and condensed refrigerant flow to a purge separator where they are separated; the liquid refrigerant returns to the system through an evaporator while the gases remain in the purge unit. When the gauge on the purge separator reaches a set pressure, it releases air and other noncondensable gases into the atmosphere along with a small amount of refrigerant.

The following checklist can help minimize refrigerant losses from the purge unit.

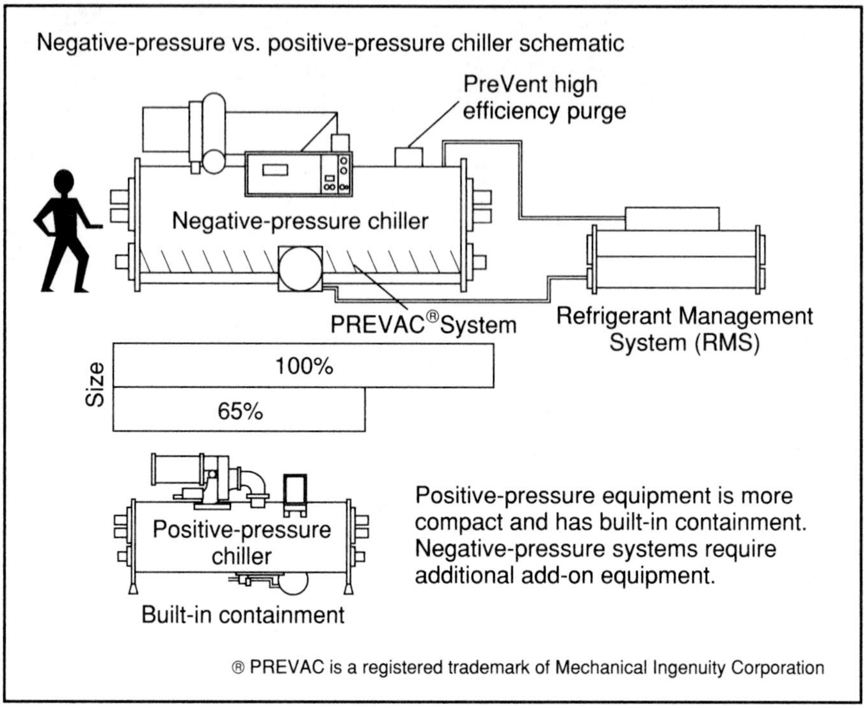

Figure 4-7 Purge unit. *Courtesy of Carrier Corporation.*

1. *Limit Purging.* One of the most effective ways to reduce refrigerant emissions is to run the purge unit only when necessary. To determine if the system needs purging, compare the temperature of the liquid in the refrigeration system condenser to the saturation temperature of the refrigeration system condenser pressure. If there is a difference of more than 2°F or 3°F, the system needs to be purged until the two temperatures agree.

2. *Relief Valve Pressure.* Care also must be taken when setting the relief device. The pressure must be set higher than the worst ambient conditions that can be experienced by the purge condenser.

   If the pressure is set too low, refrigerant will not condense out of the purge gas. The purge unit will then become an effective evacuation device, eventually pumping all the refrigerant charge out of the system.

   Consult both the pressure-temperature chart for the refrigerant being used and the purge manufacturer's manual to determine the proper set point for the relief device.

3. *Suction Control Pressure.* An improperly set suction pressure control device can also cause excessive refrigerant losses. Consult the manufacturer's manual to ensure that the control is set low enough to prevent excess refrigerant from entering the system. This in-

creases the pressure and overloads the compressor and condenser. The excess refrigerant is then vented through the relief device. This can also overload the compressor and burn out the motor.

On the other hand, the pressure control must be set high enough to keep the suction temperature above 32°F. This prevents any water in the suction line to the compressor from freezing, which could render the purge unit inoperable.

4. *Coils, Belts, Thermostats.* Condensers come in both air and liquid cooled varieties. Air cooled condensers should be checked periodically to make sure that the coils are clean. This type of condenser usually has belt driven fans. These belts should be checked to make sure that they are not worn, slipping, or broken.

With liquid cooled condensers, it is important to check the flow to determine if the proper condensing pressure is being achieved. If the liquid device uses refrigerant with a change in phase, it will usually have some type of thermostatic expansion device. This should be inspected to ensure that proper temperatures are maintained.

Any of these conditions will render the condenser inoperative and cause large quantities of refrigerant to be released to the atmosphere.

67

5. *Oil Trap.* Older purge units often have oil-lubricated compressors with a discharge oil trap. If the trap becomes plugged or the compressor pump overfills, oil can migrate into the condenser and effectively blind the heat transfer surfaces. It can also enter the purge separator and foul the return valve to the evaporator, causing the purge unit pressure to rise and discharge refrigerant through the relief valve to the atmosphere.

6. *Float Valve.* The level control device, usually a float valve, located in the bottom of the purge separator can also jam in the open position, rendering the purge unit inoperative. If the valve jams closed, the pressure in the unit will rise until the relief device opens, venting refrigerant to the atmosphere.

7. *Repair Leaks.* A purge unit that vents too frequently can also be a symptom of a leaky refrigeration system. It is important to locate and repair all leaks before putting a system back into service.

Positive-Pressure Chiller Systems

Positive pressure is a proven technology whether a vintage chiller is being replaced or a new installation is being planned. Unlike negative-pressure units, all positive-pressure equipment must be built to ASME standards. These stringent construction requirements dramatically reduce the likelihood of refrigerant losses during operation.

Because of the requirements on positive-pressure systems, non-condensables cannot enter the system during operation. Therefore, there is no need for a purge unit and no loss of refrigerant through their operation.

With positive-pressure chillers, the refrigerant charge can be contained within the system during repairs or maintenance procedures. Thus, there usually is no need for external storage tanks that risk leakage and also lengthen the amount of downtime.

Positive-pressure chillers use pressure-relief valves that release only enough refrigerant pressure to stabilize the system in an emergency. These are standard equipment on positive-pressure chillers. Negative-pressure equipment uses rupture discs that release the entire refrigerant charge to the atmosphere.

Also, the brazed joints on positive-pressure chillers are far more leak resistant than ordinary flare fittings.

Summary

◆ For CFCs to be available for several years, conservation techniques must be practiced.

◆ CFC phaseout target dates make the conservation of CFCs important. With suggested phaseout by the year 1995, rather than 2000, CFCs will be even more expensive and difficult to obtain.

◆ As of 1993, the tax on each pound of new refrigerant is $3.10 per pound, and an additional $0.45 per pound each year thereafter until complete phaseout.

◆ After July 1, 1992, the Clean Air Act prohibits the intentional venting of ozone-depleting chemicals, such as CFCs and HCFCs, from stationary air conditioning and refrigeration systems.

◆ The intentional venting of CFCs and HCFCs after July 1, 1992, carries a fine of up to $25,000 per day per offense. Each kilogram of refrigerant is counted as an offense. An Awards Provision pays up to $10,000 to anyone who reports an offender who is convicted.

◆ Only 5% of the total system charge can be vented purposely during service operations involving the refrigerant circuit.

◆ When leaks are found, refrigerant is removed from the system using a recovery unit.

◆ When leak testing a completely empty system, do not fill the system with refrigerant to purge. Pressurize the system with dry nitrogen to near operating pressure and use the bubble test. If no leaks are found, then introduce a small amount of HCFC-22 into the system and test with either a halide or electronic leak detector.

◆ This nitrogen and refrigerant mixture can then be recovered and placed in a separate cylinder and saved for future leak testing.

◆ The recovered refrigerant can be charged back into the system if it is not severely contaminated.

◆ When the system is found to be short of refrigerant, a visual check can sometimes locate the problem. The leak usually has refrigeration oil around it.

◆ An estimated 75% of the refrigerant used today is for service operations. Therefore, keeping it inside the system (containment) is the first step in CFC conservation.

◆ Negative-pressure equipment poses special problems because they tend to draw in air and moisture while in operation, and then sometimes leak refrigerant out during the idle period.

◆ Most refrigerant losses from operating or idle equipment can be eliminated by the addition of aftermarket devices such as high-efficiency purges, vacuum preventing devices, and pressure relief valves.

◆ Purge units can be a major cause of refrigerant emissions if they do not operate properly.

◆ A purge unit that purges too frequently can be a symptom of a leaky refrigeration system.

69

◆ Because of the requirements on positive-pressure systems, non-condensables cannot enter the system during operation. Therefore, there is no need for a purge unit and no loss of refrigerant through its operation.

◆ Positive-pressure chillers have pressure-relief valves that release only enough pressure to stabilize the system in an emergency.

◆ Key Terms ◆

ASME (American Society of Mechanical Engineers)

LAL (lowest achievable level)

RMS (refrigerant management system)

Review Questions

Essay

1. What is the suggested phaseout date of CFCs?

2. What will be the Federal Tax on CFC refrigerants by the year 1993?

3. Who is to be required to certify that all leaks in a refrigeration system are repaired?

4. After July 1, 1992, how much refrigerant intentionally vented is counted as an offense?

5. After July 1, 1992, how much is the maximum fine for intentional venting of refrigerant?

6. For refrigerant conservation, what is more important now than ever?

7. How is the refrigerant removed from the system after locating leaks?

8. When using dry nitrogen and HCFC for leak detection, how should the mixture be removed from the unit?

9. What should be done to the dry nitrogen/HCFC mixture before using it again?

10. What should be done to the refrigerant that is removed from a system before it is recharged back into the system?

True-False

11. The nitrogen and HCFC-22 mixture used for leak testing purposes should be purged after leak testing.

12. A refrigerant leak cannot be found by visual inspection.

13. The majority of CFCs used today are used during service operations.

14. Negative-pressure systems pose special problems for refrigerant containment and conservation.

15. High-efficiency purge systems can help prevent refrigerant leaking from negative-pressure systems.

Fill-In-The-Blank

16. A properly designed _____ _____ _____ recovers the refrigerant (including the vapor) of a negative-pressure chiller, recycles it, then returns it to the chiller.

17. _____ _____ are a small part of chillers, but they can be a major source of refrigerant emissions if they do not operate properly.

18. A purge unit that vents too _____ can also be a symptom of a leaky refrigeration system.

19. Positive-pressure systems are not equipped with a _____ _____, a major cause of refrigerant leakage.

20. Positive-pressure chillers use _____ _____ valves that release only enough refrigerant to stabilize the system in an emergency.

Chapter 5

Refrigerant Recovery, Recycling, and Reclaiming

Objectives

After completion of this chapter you should:

◆ *understand more about refrigerant recovery, recycling, and reclaiming.*

◆ *know more of the procedures used for leak testing.*

◆ *know some of the methods used for refrigerant containment.*

When discussing the CFC issue with regard to operations involving refrigerants, several definitions should be understood. The following are industry accepted definitions. These processes may be done with either recycling or recovery equipment. Refrigerant recovery, recycling, and reclaiming equipment must be certified by the EPA.

Recovery

Refrigerant recovery is the removal of refrigerant in any condition from a system and storage of it in an external container without necessarily testing or processing it in any way.

Refrigeration that has been taken from a system and has not left the job site can be placed back into the system without being treated. However, it would be well to clean the refrigerant to avoid future problems

71

resulting from the use of contaminated refrigerant. The use of a Totaltest refrigerant tester indicates if the refrigerant is contaminated with moisture or acid. It will not indicate the presence of other contaminants or if more than one refrigerant is present. Refrigerant that has been taken from the job site must be reclaimed to meet ARI (Air Conditioning and Refrigeration Institute) Standard 700–88, Table 5-1.

Recycle

To recycle means to clean the refrigerant for reuse by oil separation and single or multiple passes through devices, such as replaceable core filter-driers, which reduce moisture, acidity, and particulate matter. This term usually applies to procedures implemented at the field job site or at a local service shop.

72

	REFRIGERANTS								
	R11	R12	R13	R22	R113	R114	R500	R502	R503
PHYSICAL PROPERTIES Boiling point F @ 29.92 in. Hg	74.9 (23.8)	−21.6 (−29.8)	−114.6 (−81.4)	−41.4 (−40.8)	117.6 (47.6)	38.8 (3.8)	−28.3 (−33.5)	−49.8 (−45.4)	−127.6 (−88.7)
Boiling range °F for 5% to 85% by volume distilled	0.5	0.5	0.9	0.5	0.5	0.5	0.9	0.9	0.9
VAPOR PHASE CONTAMINANTS Air and other non-condensables (in filled container) Max. % by volume	—	1.5	1.5	1.5	—	1.5	1.5	1.5	1.5
LIQUID PHASE CONTAMINANTS Water— ppm by weight	10	10	10	10	10	10	10	10	10
Chloride ion— no turbidity to pass by test	pass	pass	pass	pass	pass	pass	pass	pass	pass
Acidity— Max. ppm by weight	1.0	1.0	1.0	1.0	1.0	1.0	1.0	1.0	1.0
High boiling residues— Max. % by volume	0.01	0.01	0.05	0.01	0.08	0.01	0.05	0.01	0.01
Particulates/Solids— visually clean to pass	pass	pass	pass	pass	pass	pass	pass	pass	pass
Other refrigerants— Max. % by weight	0.5	0.5	0.5	0.5	0.5	0.5	0.5	0.5	0.5

Table 5-1 Physical properties of fluorocarbon refrigerants and maximum contaminant levels. *Courtesy of Air Conditioning and Refrigeration Institute.*

Refrigerant that is contaminated must be recycled to remove the contaminants and oil. If contaminated refrigerant is recharged back into a system, future problems will result.

Reclaim

To reclaim a refrigerant is to reprocess the refrigerant to a new product specifications by means that may include distillation. Reclaim requires chemical analysis of the refrigerant to determine that appropriate product specifications are met. The term usually implies the use of procedures available only at a reprocessing or manufacturing facility.

It should be noted that "new product specifications" means that the refrigerant must meet ARI Standard 700-88. This requires that the refrigerant be chemically analyzed to assure that it has been truly reclaimed. It does not mean that refrigerant that has been passed through filter-driers is reclaimed.

Options

73

Several options can be taken in refrigerant operations. These will be discussed in the order that the refrigerant normally is taken from the system.

Recover and Reuse, No Processing

When the refrigerant has been removed from a system that is not contaminated, there is no need to recycle or reclaim it. The repair or the desired maintenance can be performed and the refrigerant replaced without processing. Proper procedures for removing refrigerant must be observed to prevent contamination. The service technician, or the person removing the refrigerant, must be familiar with the proper procedures and use of the recovery unit.

First, make certain that the cylinder into which the refrigerant is being placed is not contaminated and is intended for use with recovered refrigerant. This may be accomplished by completely removing any refrigerant from the cylinder and evacuating it before pumping the recovered refrigerant into it. It may be a good practice to initiate a routine in which the technician checks the quality of the refrigerant on every job and keeps a record of the results. This procedure aids in future work on the unit and determines if the system is becoming contaminated. The practice also allows the technician to check routinely the refrigerant quality to prevent problems before they occur. A good refrigerant quality check can be made with devices such as the Carrier Totaltest Kit.

The practice of recovery and reuse of refrigerant without any type of processing has several liabilities connected to it. The contractor assumes the liability that the refrigerant is in good condition, that all EPA regulations have been complied with, that possible warranty problems are assessed with the equipment manufacturer, that no CFCs have been lost, as well as other considerations.

On-Site Recovery and Recycling

When the refrigerant to be removed from a system is contaminated, it must be recycled to remove the contaminants before being recharged into the unit. This procedure is best suited to situations in which the quantity of refrigerant is small. The recovery and recycling equipment must be certified for refrigerant. This is a popular procedure in the automotive air conditioning, and domestic refrigeration and air conditioning trades.

Among the considerations for on-site recovery and recycling are: the contractor assumes liability for the quality of the refrigerant; it is sometimes difficult to meet all the ARI 700-88 specifications with on-site recycling; there may be possible equipment warranty problems; and there must be no loss of CFCs. One advantage of this method over the recovery without processing method is that the refrigerant is of better quality than when simply recovered and reused.

Recovery and Off-Site Reclaiming

When the refrigerant removed from a system is so contaminated that it is not safe to recycle it, the best solution is to reclaim it at an off-site facility. Remember: if the refrigerant is removed from the original site, it must be reclaimed before further use. The owners of some systems, especially very large or specialized systems, may require that the refrigerant be reclaimed before charging it back into the system. When the refrigerant is reclaimed it is refined to meet ARI 700-88 specifications. Reclaiming facilities reclaim the refrigerant and then have a certified laboratory verify that the refrigerant meets the necessary standards before it is sold again.

Some reclaim facilities furnish the necessary cylinders and other support that may be needed to meet local, state, and federal codes for moving contaminated refrigerant.

When the reclaim method is chosen, there are many advantages to the service technician or the service company. The reclaim facility assumes full responsibility for the refrigerant quality; there is proof that the regulations have been complied with; there are no taxes on reclaimed refrigerant; and the refrigerant is available for use in the future.

Recover and Destroy

When the refrigerant to be recovered is badly contaminated or mixed with other refrigerants, it cannot be reclaimed. The only alternative is to destroy it. This is a very difficult process because CFCs are so stable that they are not easily destroyed.

In most instances, when different refrigerants are mixed, it is extremely difficult to separate them. Some reclaim facilities are equipped to separate different types of refrigerants, but this is a very expensive process.

The only method to destroy refrigerants effectively is incineration. This is an expensive operation because the process must contain the released fluorine. Contaminated refrigerant always must be sent to a facility authorized for handling it. The oil in contaminated systems also is contaminated and must be disposed of properly.

Several advantages and disadvantages are involved when the recovery and destroy method is chosen. New refrigerant must be purchased, and tax is levied on new refrigerant; the refrigerant is lost and unavailable for future use; shipping costs usually are incurred; and the cost of disposal is high.

Refrigerant Specifications

75

There are basically two specifications used for refrigerants. The automotive industry uses the SAE J1991 specifications and the remainder of the industry uses the ARI Standard 700-88 Specifications for Fluorocarbon Refrigerants Standard.

Summary

- ◆ The recovery of a refrigerant is the removal of refrigerant in any condition from a system and storage of it in an external container without necessarily testing or processing it in any way.

- ◆ Refrigerant that has been taken from a system and has not left the job site can be placed back into the system without being treated. The refrigerant should be cleaned to avoid future problems.

- ◆ The use of a Totaltest refrigerant tester indicates whether the refrigerant is contaminated with moisture or acid.

- ◆ To recycle means to clean the refrigerant for reuse by oil separation and single or multiple passes through devices such as replaceable core filter-driers, which reduce moisture, acidity, and particulate matter.

◆ To reclaim refrigerant is to process it to new product specifications by means that include distillation. The term usually implies use of processes or procedures available only at a reprocessing or manufacturing facility.

◆ When refrigerant has been removed from a system that is not contaminated, there is no need to recycle or reclaim it.

◆ When recovering refrigerant, make certain the cylinder into which the refrigerant is being placed is not contaminated and is intended for use with recovered refrigerant.

◆ It may be good practice for technicians to check the quality of the refrigerant on every job and keep records of the results. This procedure aids in determining if the system is becoming contaminated.

◆ The practice of recovery and reuse of refrigerant without any type of processing has several liabilities associated with it: the contractor assumes the liability that the refrigerant is in good condition and that all of the EPA regulations have been complied with; possible warranty problems are assessed with the equipment manufacturer; and no CFCs may be lost. This method is best suited when small quantities of refrigerant are involved.

◆ The recovery and recycling equipment must be certified for refrigerant.

◆ When the refrigerant removed form a system is so contaminated that it is not safe to recycle it, it should be reclaimed at an off-site facility.

◆ Among the advantages of the reclaiming method are that the reclaim facility assumes full responsibility for refrigerant quality; there is proof that regulations have been complied with; there are no taxes on reclaimed refrigerant; and the refrigerant is available for reuse in the future.

◆ When the refrigerant to be recovered is badly contaminated or mixed with other refrigerants and cannot be reclaimed, the only alternative is to destroy it. This is a very difficult process because CFCs are so stable that they are not easily destroyed.

◆ Incineration is the only method to effectively destroy mixed refrigerants. The operation is expensive because the process must contain the released flourine.

76

◆ Key Terms ◆

reclaim

recovery

recycle

Totaltest tester

Review Questions

Essay

1. What should be done to recovered refrigerant before placing it back into the system?

2. If refrigerant is taken from the job site, what must be done before reusing it in another owner's unit?

3. What device can be used to determine if a refrigerant contains acid or moisture?

4. What is refrigerant cleaning, but not necessarily to ARI 700-88 specifications, known as?

5. What must be done to refrigerant that changes ownership?

Fill-in-the-Blank

6. A Totaltest tester will not determine if _____ are present in a refrigerant.

7. To clean refrigerant to new product specifications is known as _____.

8. When recovering refrigerant, the first thing to do is to make certain that the cylinder into which the refrigerant is being placed is not contaminated and is for _____ refrigerant.

9. When the refrigerant to be removed from a system is found to be contaminated, it must be _____ to remove the contaminants.

10. _____ the refrigerant is very popular in the automotive air conditioning industry.

True-False

11. Keeping a record of the quality of refrigerant in a system on every job aids in determining if the system is becoming contaminated.

12. The practice of recovery and reuse of refrigerant without any type of processing has several liabilities.

13. The customer takes the liability for recycled or reclaimed refrigerant.

14. When refrigerant is to be used in a unit owned by an owner different from the first unit it must be recycled before it can be reused.

15. The reclaim facility assumes full responsibility for any refrigerant reclaimed by it.

Multiple Choice

16. Before recharging refrigerant back into a system, it should be
    a. heated.
    b. cooled.
    c. recycled.
    d. reclaimed.

17. To reclaim refrigerant means to
    a. clean the refrigerant.
    b. pass it through filter-driers.
    c. clean it to SAE J1990 specifications.
    d. clean it to ARI 700-88 specifications.

18. Destroying contaminated CFCs is very difficult because
    a. they are so stable.
    b. they are difficult to find.
    c. no one wants to do it.
    d. the owner wants to keep them.

19. The only method to effectively destroy CFCs is
    a. releasing them through water.
    b. incineration.
    c. placing them in a clean cylinder for several months.
    d. circulating them through filter-driers.

20. To recover refrigerant means to
    a. remove it from the system.
    b. place the refrigerant in a DOT 39 cylinder.
    c. pass the refrigerant through filter-driers.
    d. clean it to ARI 700-88 specifications.

Chapter 6

Methods of Refrigerant Recovery and Recycling

Objectives

After completion of this chapter you should:

◆ *know the different methods of refrigerant recovery and recycling.*

◆ *be more familiar with the safety procedures for use in handling refrigerants.*

◆ *know more about handling refrigerant cylinders.*

When recovering refrigerant from a system it is important to take precautions to protect both the personnel as well as the equipment. Certain procedures must be followed to prevent accidents. The following are some of the more important safety practices.

Health Hazards

The following information describes safety considerations associated with fluorocarbon refrigerants. Before filling refrigerant cylinders or handling any fluorocarbon product, product data sheets for the specific product should be obtained and reviewed thoroughly.

   CFC products usually are low in oral and inhalation toxicity; however, they possess properties that can pose significant health hazards under certain circumstances. Exposure to fluorocarbons above the rec-

79

ommended exposure levels (threshold limit values or TLVs) can result in drowsiness and loss of concentration. Experiments with laboratory animals demonstrate that cardiac arrhythmias can be induced by fluorocarbon levels of 20,000 to 100,000 parts per million. Cases of fatal cardiac arrhythmias in humans accidentally exposed to these high levels also have been reported.

Studies with laboratory animals also have demonstrated that fluorocarbons can sensitize the heart muscle to epinephrine (adrenaline). This may lower the threshold at which adverse affects (cardiac arrhythmia) may occur when accidental overexposure occurs during strenuous physical activity. This finding dictates that epinephrine not be administered as a medical treatment for any overexposure to these products.

CFCs are heavier than air and can displace the air from enclosed or semienclosed areas. Therefore suffocation is a potential hazard. Do not fill refrigerant reclamation shipping containers in low or enclosed areas without proper ventilation.

Skin exposure to CFC products can result in irritation, primarily due to the fluorocarbon's defatting action. Skin or eye contact can result in frostbite, primarily from liquefied or compressed gases. Use personal protection equipment, such as side shield glasses, gloves, safety shoes, and a hard hat when filling and handling reclamation containers.

Do not expose CFC products to sparks, open flame, lit cigarettes, or hot surfaces (250°F/121.1°C). Many fluorocarbon materials decompose if exposed to temperatures above this level or open flames in the presence of air. Thermal decomposition products include hydrochloric acid and hydrofluoric acid vapors, both of which are irritating and toxic. In addition, carbonyl halides (e.g., phosgene) can theoretically be produced during decomposition. These materials are toxic. Do not apply open flame to or heat reclamation containers above 125°F (52°C).

The following suggestions will help in the safe handling of refrigerant.
1. Always read the product label and the product Safety Data Sheet.
2. Always use adequate ventilation.
3. Never expose these products to flames, sparks, or hot surfaces.
4. **Physicians** should not use epinephrine to treat overexposure.

Cylinders

Cylinders must always be filled by weight, never more than 80% full at 70°F. Consequently, a scale is essential for safe cylinder filling. The maximum allowable of any refrigerant that can be recovered into a 122-pound w.c. cylinder is 100 pounds. The gross weight (cylinder plus contents) of a filled cylinder must not exceed 150 pounds. A proper gauge set manifold also is required for filling to assure that the charging pressure does not exceed 300 psig.

80

Specifications for Containers for Recovered Fluorocarbon Refrigerants

ARI Guideline K is the guide for cylinders used for recovering CFC refrigerants. The containers, color code, and filling procedure follow. For complete coverage of these specifications, see ARI Guideline K.

Containers

Cylinders for recovered fluorocarbon refrigerants CFC-12, CFC-22, CFC-114, CFC-500, and CFC-502 must comply with the following:

- United States Department of Transportation (DOT) specification packaging, see Title 49 CFR (Code of Federal Regulations), and have a service pressure rating of not less than 260 psig.
  Note: Previously filled DOT specification 39 *disposable* cylinders shall not be used for the storage and/or transportation of recovered fluorocarbon refrigerants. Federal law forbids transportation of specification 39 disposable cylinders if refilled. Penalty for violating this requirement is up to $25,000 fine and five years' imprisonment [Title 49 U.S.C. (United States Code) Sec. 1809].
- Valves shall comply with Compressed Gas Association Standard V-1, "Compressed Gas Cylinder Valve Outlet and Inlet Connections."
- Safety relief valves shall comply with "Pressure Relief Device Standard Part 1—Cylinders for Compressed Gases," Compressed Gas Association Pamphlet S-11.

Ton Tanks for Recovered Fluorocarbon Refrigerants CFC-12, CFC-22, CFC-114, CFC-500, and CFC-502 shall comply with the following:

- United States Department of Transportation (DOT) specification 106A500X or 110A500W as detailed in Title 49 CFR, Section 179.300.
- Valves shall comply with Compressed Gas Association Standard V-1, "Compressed Gas Cylinder Valve Outlet and Inlet Connections."
- Safety relief devices shall comply with Title 49 CFR, Section 179.300-15.

Containers for Recovered Fluorocarbon Refrigerants CFC-11 and CFC-113 steel drums shall comply with Title 49 CFR, DOT specifications 17C or 17E, as set forth in Title 49 CFR 178.115, 178.116.

Containers that originally contained new refrigerant CFC-11 or refrigerant CFC-113 (excluding those originally used for cleaning agents) may be used, provided the following conditions are met:

- The drums shall be inspected internally and externally and found to be clean and free of dents, bulges, holes, cracks, rust, pits, creases, or other structural weaknesses.
- Closure devices, including gaskets, shall be in such condition that they comply, in all respects, with the original requirements for the drum.
- Drums that originally contained CFC-11 or CFC-113 must be made to comply with Section 6.5.3. Previous labels and markings must be removed and be replaced with new labels and markings per Section 6.3.

Cylinder Color Code

Examples of color schemes for various recovery containers follow. Depending upon the provider of the recovery container, the actual coloring may vary. However, the use of yellow in the manner described will identify the container as a recovery vessel.

Cylinders with nonremovable collars. The body shall be gray. The collar shall be yellow.

Cylinders with removable caps. The body shall be gray. The shoulder and the cap shall be yellow.

Drums. The drum shall be gray. The top head shall be yellow.

Tons. The body shall be gray. The ends and chimes shall be yellow.

82

Filling Procedure

Cylinders and Ton Tanks. Do not fill if the present date is more than five years past the test date on the container. The test date is stamped on the shoulder or collar of cylinders and on the valve end chime of ton tanks and appear as follows:

A1

12   89

32

This indicates the cylinder was retested in December of 1989 by retester number A123.

Cylinders and ton tanks shall be weighed during filling to ensure user safety. Maximum gross weight is indicated on the side of the cylinder or ton tank and shall never be exceeded.

Cylinders and ton tanks shall be checked for leaks prior to shipment. Leaking cylinders and ton tanks must not be shipped.

A procedural checklist is included with each cylinder and reproduced here. Read it prior to filling and use it as a guide during filling to assure that all steps are taken.

Prior to filling this recovery cylinder, identify the refrigerant to be recovered and make certain that the cylinder is marked and labeled for that refrigerant. Remove the "Empty" tag and apply the product label.

Do not mix different refrigerant gases in the same cylinder. Verify the serial number against the Cylinder Control tag. Read all labels.

Make certain that the cylinder test date has not expired. ***Do not fill*** if the present date is more than 5 years past the most recent test date. The test date appears on the shoulder of the cylinder.

Inspect the cylinder for signs of damage, such as dents, gouges, and corrosion. *Do not fill* damaged cylinders. Inspect the valve for damage and ease of operation. *Do not fill* cylinders with damaged valves.

Determine the maximum allowable gross weight—this will appear in large letters on the side of the cylinder. Maximum allowable gross weight of a 122-pound w.c. cylinder is 150 pounds.

Be sure to have an accurate scale suitable for weighing the cylinder and contents, a proper gauge set manifold, and proper hoses and connectors.

Be sure that the cylinder is freestanding on the scale with no restriction of free movement caused by the hoses, connections, etc.

Monitor the pressure during the filling process carefully. ***Do not exceed*** 300 psig.

Monitor the gross weight during filling to prevent overfilling. An overfilled cylinder can rupture, resulting in serious injury or death. The maximum gross weight is 15 pounds (122 pounds w.c.).

Shut off the valve if the marked gross weight or 300 psig pressure is reached. ***Do not overfill.***

After recovery, close the cylinder valve securely.

Leak check. ***Do not ship a leaking cylinder.***

Apply the outlet cap.

Enter the gross weight and shipper identification on the Cylinder Control tag. Sign the certification and insert the tag in the plastic envelope. Attach it to the cylinder valve.

Apply the steel valve protector cover. Make certain that the Cylinder Control tag is enclosed.

Attach the DOT diamond tag to the valve cover.

WARNING! DO NOT OVERFILL: An overfilled cylinder can rupture, resulting in serious personal injury or death.

83

Drums

Only recovered refrigerant CFC-11 shall be placed into a drum that previously contained new refrigerant CFC-11. Only recovered refrigerant CFC-113 shall be placed into a drum that previously contained new refrigerant CFC-113. Drums that originally contained CFC-113 for use as a cleaning agent shall not be used.

Drums shall be filled to allow a vapor space equal to at least 10% of the drum height between the top of the liquid and the bottom of the drum top.

Drums shall be sealed by wrench-tightening the closure devices until the gaskets are firmly seated.

Drums shall be checked for leakage prior to shipment. Leaking drums must not be shipped.

Drums may be filled either by weight or volume. When filling by weight, do not exceed the gross weight (drum plus contents) marked on the drum. When filling by volume, allow a vapor space equal to at least 10% of the height between the top of the liquid and the bottom of the drum top.

Make certain to follow these procedures, which are included with each drum, when filling.

Remove the "Empty" label.

Apply the "Recovered Refrigerant II" product label to the drum on the side opposite the stenciled marking.

Open the plug slowly, between fills, to relieve possible internal pressure.

Fill to allow a vapor space of at least 10% of the drum height or to the maximum gross weight marked on the side of the drum.

Wrench tighten both plugs and check for leaks.

Weigh drum and enter the gross weight and the shipper identification on the "Drum Control Tag." Sign the certification and insert the tag in the plastic envelope attached to the drum head. Make certain that the serial numbers on the Drum Control Tag match the serial number on the label.

Storage and Handling

Store recovery cylinders and drums in a dry, ventilated warehouse or other enclosed area away from heat, flame, corrosive chemicals or fumes, and explosives. Keep the containers away from direct sunlight, particularly in warm weather. These refrigerants expand significantly when heated so that some head space must be left in the container to allow for this expansion as the containers become warm. If overfilled, cylinders and drums may become liquid full. Once liquid full any further rise in the temperature of a closed container can result in an abrupt rise in liquid pressure to extremely high levels—enough to cause the container to burst violently, resulting in serious personal injury or death. **Never allow a cylinder to get warmer than 125°F (52°C).**

Recovery cylinders and drums should always be raised above dirt or damp floors to prevent rusting. Use a platform or parallel rails. All containers must be secured in place by means of a rack, chain, or rope, so they cannot tip, roll, or accidentally strike each other or any other object. The storage area should be remote from corrosive chemicals or fumes, to avoid attack on the cylinders and damage to threaded areas of the valve and cylinder.

To minimize damage and exposure to injury, observe the following good practices when handling cylinders and drums, and read the product labels.

Cylinders

- Keep the outlet cap on the valve outlet and the valve hood securely screwed onto the neck of returnable cylinder at all times except when recovering refrigerant.
- Never drop a cylinder.
- Keep returnable cylinder secured in the upright position.
- Never hit a cylinder with a hammer or any other object.
- Never apply live steam or direct flame to a cylinder.
- Do not lift the cylinder by the valve cover or the valve.
- Never remove the valve from the cylinder.
- Do not attempt to repair the valve.
- Do not tamper with the safety device.
- Do not remove or attempt to alter any permanent cylinder markings. (It is illegal to do so.)
- Take extreme care not to dent, cut, or scratch the cylinder or valve.
- Protect the cylinders from moisture, salt or corrosive chemicals, or atmosphere, in any form.
- Always open the valve slowly.
- Close the valve slowly.
- Close the valve tightly after use.
- Do not attempt to use a cylinder in a rusted or otherwise deteriorated condition—contact the appropriate personnel for disposal.

85

Drums

- Do not use pressure to empty the drum.
- Before moving the drum after each use, be sure that the plugs are secured tightly.

Labeling

Section 611 of the Clean Air Act requires mandatory warning labels on all products and containers of ozone-depleting substances introduced into interstate commerce as early as May 15, 1993.

Containers Containing Class I or Class II Substances and Products Containing Class I Substances

Effective 30 months after the enactment of the Clean Air Act Amendments of 1990, no container in which Class I or Class II substances are stored or transported, and no product containing a Class I substance shall be introduced into interstate commerce unless it bears a clearly legible and conspicuous label stating:

> **Warning: Contains** (insert name of substance), a substance which harms public health and environment by destroying ozone in the upper atmosphere.

Follow the instructions for filling out the label for specific names and UN identification numbers of the refrigerant it contains.

CFC-11, CFC-113, and CFC-114 are not considered hazardous by DOT. Therefore, no labeling is required on containers for these substances.

86

Refrigerant Recovery Methods

Section 608 of the National Recycling and Emission Reduction Program states that regulations under subsection (a) shall establish standards and requirements for the safe disposal of Class I and Class II substances. Such regulations shall include each of the following:

1. Requirements that Class I or Class II substances contained in bulk in appliances, machines, or other goods shall be removed from each such appliances, machines or other goods prior to the disposal of such items or their delivery for recycling.
2. Requirements that any appliance, machine, or other goods containing a Class I or a Class II substance in bulk shall not be manufactured, sold, or distributed in interstate commerce or offered for sale or distribution in interstate commerce unless it is equipped with a servicing aperture or an equally effective design feature, that facilitates the recapture of such substance during service and repair or disposal of such item.

Basically, recovery means that the refrigerant is to be taken from a system to be repaired and placed in a storage cylinder until repairs are completed. This can be accomplished in two ways: liquid recovery and vapor recovery. Both methods draw the refrigerant into a storage cylinder.

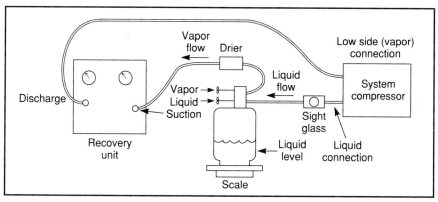

Figure 6-1   Liquid recovery connections.

Liquid Recovery

It is usually much faster to recover liquid than vapor. Thus, when possible, remove the liquid first and then the vapor. To use the liquid recovery method, use the connections shown in Figure 6-1.

Make certain that the liquid refrigerant does not enter the compressor on this type of unit to prevent damage to the compressor. In this method the liquid refrigerant flows into the cylinder because of the pressure difference. The suction of the recycle unit lowers the pressure in the cylinder, and the discharge from the unit causes a higher pressure on the liquid, causing it to flow into the cylinder. Notice the drier in the vapor line to the recovery unit. This will aid in cleaning the refrigerant.

87

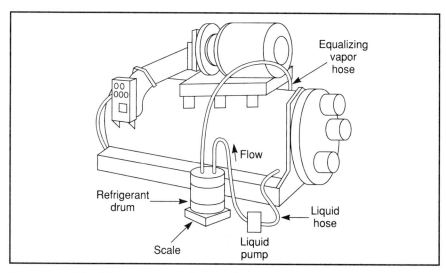

Figure 6-2   Liquid pump connections for large amounts of refrigerant.

Liquid Refrigerant Pump. A liquid refrigerant pump is used when removing large quantities of refrigerant from a system, such as a chiller or some large industrial unit. Connect the liquid pump in the line as shown in Figure 6-2.

After the liquid has been removed, the vapor can then be removed by a regular recovery unit. This method is much faster than trying to vaporize the liquid and take it out in the vapor form. Be sure to use a liquid refrigerant pump to prevent damaging the pump.

Vapor Recovery

To recover the refrigerant vapor from a system to be repaired, connect the recovery unit as shown in Figure 6-3.

The vapor is drawn into the recovery unit, compressed, condensed, and then forced into the cylinder. Notice that there is only one connection to the system. This method is suited for a smaller system where there is very little liquid refrigerant to be removed. Notice the drier in the vapor line to the recovery unit. This will aid in cleaning the refrigerant.

88

Refrigerant Recovery/Recycle Methods

The refrigerant recovery methods previously discussed are the same regardless of the purpose for the recovery. Here the recycle requirements and methods are discussed.

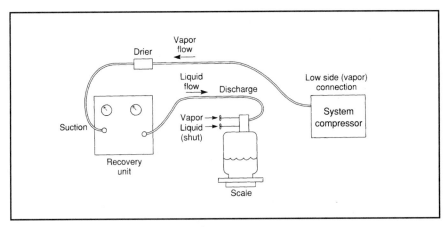

Figure 6-3   Vapor recovery connections.

Recycle Requirements

The refrigerant used in any refrigeration or air conditioning system must be recovered after July 1, 1992. What is to happen after it has been recovered is usually decided at the time of recovery or during the recovery process. The refrigerant can either be returned directly to the system from where it came; it can be recycled for use in the same system; or it can be delivered to a reclaim facility for reclaiming or disposal.

Refrigerant that has been recovered from a system in order to perform routine service or a major overhaul usually can be returned directly to the same unit with no recycling procedures if the refrigerant is not contaminated. If there is any question about the purity of the refrigerant, it must be analyzed for moisture, acids, high boiling-point residues, and other types of contaminants listed in ARI Standard 700-88 and any required steps taken to either bring the refrigerant back within specifications or destroy it. The proper evacuation and charging procedures are discussed later in this chapter.

When refrigerant contamination, moisture, motor burnout, or mechanical failure is the cause for the unit being out of service, the refrigerant must be analyzed for contaminants and recycled before it is charged back into the system. The recycling instructions for the particular recover/recycle unit being used should be followed. The general charging procedure is as follows:

1. Leak test the system. Pressurize the system with dry nitrogen to near the operating pressure and let it set idle for a period of time and observe for any drop in pressure in the system. If there is a loss of pressure leak test any suspected joints or other places with a liquid plastic leak detector available at most supply houses. Observe the suspected area for about 5 seconds. If bubbles appear there is a leak that must be repaired before the system is recharged with refrigerant. Check all suspected areas before releasing the pressure so that all leaks can be repaired without pressurizing the system again. If no leaks are found with this method, use the following method.

2. Release some of the nitrogen pressure, introduce HCFC-22 vapor into the system, and increase the pressure to near cylinder pressure. Then use either the halide torch, the electronic leak detector, or some other tool to test for leaks with refrigerant vapor in the system. Check all suspected areas before releasing the pressure so that all the leaks can be repaired at one time. Do not purge this nitrogen/HCFC-22 mixture. Remove it and place in a special cylinder for future leak testing use.

3. Install new liquid line filter-driers and any required suction line filter-driers. Use the proper precautions to prevent contamination of either the system or the filter-driers.

89

4. After it has been determined that there are no leaks in the system, evacuate the unit by connecting the vacuum pump to the unit in the usual manner. Proper evacuation will remove moisture and noncondensables from the system. Evacuate the system to approximately 1000 microns of mercury. If the system is not capable of holding a 2500 micron or less vacuum, it must be leak tested and all leaks repaired.

   There are two main methods to evacuate a refrigeration system: the deep vacuum method and the triple evacuation method.

   Deep Vacuum Method. Draw a vacuum to an absolute pressure of less than 100 microns mercury. Close off all the hand valves and allow the unit to stand idle for several hours, if possible. This is to make sure that the system will hold a vacuum of at least 2500 microns. On larger systems, 24 hours may be required to determine this.

   Triple Evacuation Method. Draw a vacuum of about 27 inches of mercury. Break this vacuum with HCFC-22 to about zero gauge pressure and allow the system to stand idle for at least one hour. Reevacuate the system and repeat this process two or three times. The HCFC-22 used for this procedure should be recovered, not purged to the atmosphere, and saved in a cylinder for future leak testing.

   Note: This method is not recommended for moisture removal or for low-pressure systems.

   Do not evacuate the system so fast that any moisture inside it will freeze. Ice will remain in the system even after a vacuum has been on the system for several hours. When the ice melts, the resulting moisture can cause possible damage to the system. When a satisfactory vacuum has been accomplished, the system can then be recharged.

5. Recharge the system with the proper amount of refrigerant. If the system is not to be fully charged at this time, introduce a holding charge of the proper type refrigerant to bring the system to a positive pressure.

Recycling for Reuse

When the refrigerant is to be recovered from a system that is contaminated or a motor burnout, the refrigerant must be analyzed for contaminants and recycled before it is used again. This process may involve only refrigerant filtering and drying, or it may require that the oil also be removed from it.

When the refrigerant is to be taken off-site, even for use in a similar system, it must be recycled to remove any possible contaminants. The equipment to do this must be certified as recycling equipment. The

90

refrigerant must meet the requirements of ARI Standard 700-88 for reuse in another system, except a system that is owned by the same person. Then the recovered/recycled refrigerant can be used in another system.

Refrigerants that are mixed with other gases must be reclaimed. Refrigerants that have been mixed with nitrogen for leak testing purposes can be placed in a container and used for future leak testing. It is a good idea to clean and dry this mixture before introducing it into another system to prevent contamination. In any case, do not purge it to the atmosphere.

Recovery, Recycling, and/or Reclaim Equipment Standards

ARI Standard 740 is the standard to which these units must operate. The General Equipment requirements of the standard follow:

The equipment manufacturer shall provide operating instructions, necessary maintenance procedures, and source information for replacement parts and repair.

91

The equipment shall reliably indicate when the filter-drier(s) needs replacement if this method is used.

The equipment shall either automatically purge noncondensables if the acceptable level is exceeded or alert the operator that the noncondensable level has been exceeded.

The equipment's refrigerant loss due to noncondensable purging shall not exceed 5% by weight of total recovered refrigerant.

Internal hose assemblies shall not exceed a permeation rate of 12 pounds mass per square foot (5.8 g/cm²) of internal surface per year at a temperature of 120°F (48.8°C) for any designated refrigerant.

The equipment shall be capable of operation to the specifications in ambient temperatures of 50°F to 104°F (10°C to 40°C).

Equipment specified to operate within a controlled temperature range must be evaluated only within that range.

Exemptions: Equipment intended for a single professional operator and backed by chemical analysis shall be exempt from the preceding requirements except "The equipment's refrigerant loss due to noncondensable purging shall not exceed 5% by weight of the total recovered refrigerant."

Contaminated Refrigerants

The standard contaminated refrigerant sample shall have the contents as specified in Table 6-1.

	R11	R12	R13	R22	R113	R114	R500	R502	R503
Moisture Content: PPM by Weight of Pure refrigerant	100	80	30	200	100	85	200	200	30
Particulate Content: PPM by Weight of Pure Refrigerant Characterized by[1]	80	80	80	80	80	80	80	80	80
Acid Content: PPM by Weight of Pure Refrigerant—mg KCH per kg Refrig.) Characterized by[2]	500	100	N/A	500	400	200	100	100	N/A
Mineral Oil Content: % by Weight of Pure Refrigerant	20	5	N/A	5	20	20	5	5	N/A
Viscosity (SUS)	300	150	300	300	300	300	150	150	300
Non Condensable Gases Air Content % Volume	N/A	3	3	3	N/A	3	3	3	3

[1]Particulate content shall consist of inert material and shall comply with particulate requirements in ASHRAE Standard 63.2, "Method of Testing the Filtration Capacity of Refrigerant Liquid Line Filters and Filter Driers."

[2]Acid consists of 60% oleic acid and 40% hydrochloric acid on a total acid number basis.

92

Table 6-1 Standard contaminated refrigerant sample. *Courtesy of Air Conditioning and Refrigeration Institute.*

Summary

◆ Before filling refrigerant cylinders or handling any fluorocarbon product, it is suggested that a copy of the product data sheets for the specific product be obtained and thoroughly reviewed.

◆ CFC products are usually low in oral and inhalation toxicity, however, they possess properties that can pose significant health hazards under certain circumstances. Exposure to fluorocarbons above the recommended exposure levels (threshold limit value) can possibly result in drowsiness and loss of concentration.

◆ Past experiments with laboratory animals have demonstrated that cardiac arrhythmias can be induced by fluorocarbon levels of 20,000 to 100,000 parts per million.

◆ It has also been demonstrated in laboratory animals that fluorocarbons can sensitize the heart muscle to epinephrine (adrenaline). This may lower the threshold limit at which adverse affects (cardiac arrythmia) may occur when accidental overexposure occurs during strenuous physical activity. This factor also dictates that epinephrine not be administered as a medical treatment for any overexposure to these products.

◆ CFCs are heavier than air and can displace the air from enclosed or semienclosed areas. This can result in a suffocation hazard. Do not fill refrigerant reclamation shipping containers in low or enclosed areas without proper ventilation.

◆ The thermal decomposition of CFCs produces hydrochloric acid and hydrofluoric acid vapors, which are irritating and toxic. In addition, carbonyl halides (phosgene) theoretically can be produced during decomposition.

◆ Refrigerant recovery cylinders always must be filled by weight, never more than 80% full at 70°F. A proper gauge manifold is required for filling to assure that the charging pressure does not exceed 300 psig.

◆ A procedural checklist is included with each cylinder. Read it prior to filling and use it as a guide during the filling process to assure that all steps are taken. The recommended procedure follows.

Prior to filling a recovery cylinder identify the type of refrigerant to be recovered and make certain that the cylinder is marked and labeled for that type refrigerant. Remove the "Empty" tag and apply the product label.

93

Do not mix different refrigerant gases in the same cylinder. Verify the serial number against the "Cylinder Control" tag. Read all labels.

Check that the cylinder test date has not expired and do not fill if the present date is more than five years past the most recent test date.

Inspect the cylinder for signs of damage, such as dents, gouges, and corrosion. Do not fill damaged cylinders. Inspect the valve for damage and ease of operation. Do not fill cylinders with damaged valves.

Determine the maximum allowable gross weight—Be sure that the cylinder is freestanding on the scale with no restriction of free movement caused by the hoses, connections, etc.

Monitor the pressure during the filling process. Do not exceed 300 psig.

Monitor the gross weight. The maximum gross weight is 150 pounds (122 pounds w.c.).

After recovery, close the cylinder valve securely, leak check, and apply the outlet cap.

Enter the gross weight and the shipper identification on the "Cylinder Control" tag. Sign the certification, insert the tag in the plastic envelope, and attach it to the cylinder valve.

Apply the steel valve protector cover. Make certain that the "Cylinder Control" tag is enclosed. Attach the DOT diamond tag to the valve cover.

◆ Refrigerant drums may be filled either by weight or volume. When filling by weight, do not exceed the gross weight (drum plus contents) marked on the drum. For filling by volume, allow a vapor space equal to at least 10% of the height of the drum.

◆ Make certain to follow these procedures, which are included with each drum, when filling:

Remove the "Empty" label.

Apply the "Recovered Refrigerant" product label to the drum on the side opposite the stenciled marking.

Open the plug slowly between fills to relieve possible internal pressure.

Fill to allow a vapor space of at least 10% of the drum height or to the maximum gross weight marked on the side of the drum.

Wrench tighten both plugs and check for leaks.

Weigh the drum and enter the gross weight and the shipper identification on the drum control tag.

Sign the certification and insert the tag in the plastic envelope attached to the drum head.

Make certain that the serial number on the "drum control" tag matches the serial number on the label.

◆ Store recovery cylinders and drums in a dry, ventilated warehouse or other enclosed area away from heat, flame, corrosive chemicals, and explosives. Never allow a container to get warmer than 125°F (52°C).

◆ Section 608 of the National Recycling and Emission Reduction Program states that: The regulations under subsection (a) shall establish standards and requirements for the safe disposal of Class I and Class II substances.

◆ It is usually much faster to recover liquid than vapor.

◆ A liquid refrigerant pump is useful when removing large quantities of refrigerant from a system such as chiller or some large industrial unit.

◆ Vapor refrigerant recovery is best suited for smaller systems containing very little liquid refrigerant.

◆ When refrigerant contamination, moisture, motor burnout, or mechanical failure is the cause for a unit being out of service, the refrigerant must be analyzed for contaminants and recycled before it is charged back into the system.

94

◆ Check all suspected areas for leaks before releasing the pressure so that all the leaks can be repaired at one time.

◆ The two evacuation methods are the deep vacuum method and the triple evacuation method.

◆ When the refrigerant is to be taken off-site, it must be recycled to remove possible contaminants.

◆ Key Terms ◆

cardiac arrhythmia

reclaim

recover

recycle

Review Questions

Essay

1. What should be reviewed before filling refrigerant recovery cylinders or handling any fluorocarbon products?

2. What should never be given to a person who has had an overexposure to fluorocarbon products?

3. Why should reclaim containers never be filled in low or enclosed areas without proper ventilation?

4. Why should CFCs never be exposed to surfaces above 250°F?

5. How are most refrigerant cylinders filled?

6. What should be done about the refrigerant before filling a recovery cylinder?

7. When recovering refrigerant what should be done to the cylinder?

8. What should never be done when recovering refrigerant?

9. Why should the cylinder test date be checked before filling?

10. What is the amount of vapor space required above the liquid in a recovery drum?

Fill-in-the Blank

11. The maximum amount that a recovery cylinder can be filled is

____ __ _____.

12. The maximum pressure allowed when filling a refrigerant recovery cylinder is ____ _____.

13. Prior to filling a recovery cylinder, remove the _____ _____ and apply the product label.

14. When filling a recovery cylinder always verify the _____ _____ against the "Cylinder Control" tag.

15. Never fill a recovery cylinder if the test date is more than _____years past.

16. A _____ cylinder should never be refilled.

17. The maximum allowable _____ _____ will appear in large letters on the side of the cylinder.

18. When filling a recovery cylinder, _____ the gross weight to prevent overfilling.

19. A _____ cylinder should never be shipped.

20. Attach the DOT diamond tag to the valve _____.

True-False

21. It is suitable to "feel" the weight when refilling a refrigerant recovery cylinder.

22. The proper procedure for filling a recovery cylinder is to fill it until the refrigerant stops going into it.

23. When filling a recovery cylinder, shut off the valve if the marked gross weight or 300 psig is reached.

24. Refrigerant drums may be filled either by weight or volume.

25. The permanent markings on a cylinder can safely be altered for different types of refrigerant.

26. Section 608 of the National Recycling and Emission Reduction Program establishes standards and requirements for Class I substances only.

27. It is faster to recover vapor refrigerant than liquid refrigerant.

28. Refilling procedures are generally furnished with recovery cylinders and drums.

29. Recovery drums can be safely filled completely full of liquid refrigerant.

30. A liquid refrigerant pump will remove both liquid and vapor refrigerant.

Multiple Choice

31. The best refrigerant recovery method for small systems is
    a. vapor.
    b. liquid.
    c. vapor and liquid.
    d. purging.

96

32. The refrigerant recovered from a system when the owner retains ownership must be
    a. reclaimed.
    b. stored at the job site.
    c. recycled for future use.
    d. destroyed.

33. When a system is contaminated by a major burnout, the refrigerant must be
    a. reclaimed.
    b. analyzed.
    c. destroyed.
    d. only recovered.

34. Leak testing for difficult leaks can be satisfactorily done by using
    a. nitrogen.
    b. dry nitrogen/HCFC-22 mixture.
    c. CFC-12.
    d. dry oxygen.

35. Two recognized evacuation procedures are
    a. single and double evacuation.
    b. double and triple evacuation.
    c. deep and triple evacuation.
    d. deep and double evacuation.

97

Commercial and Residential Stationary Air Conditioning and Refrigeration Systems

Chapter 7

Commercial Stationary Air Conditioning and Refrigeration Systems

Objectives

After completion of this chapter you should:

◆ *know how the EPA designates the license categories.*

◆ *know more about the refrigerant replacement requirements for low-pressure units.*

◆ *know more about the refrigeration replacement requirements for high-pressure units.*

◆ *be more aware of the requirements for HFC-134a.*

◆ *know more of the procedures for recovering both liquid and vapor refrigerant from the system.*

◆ *know the proper leak testing procedures used for refrigerant conservation.*

◆ *know how to service systems without purging the refrigerant when no service connections are provided.*

◆ *be more aware of the procedures required for disposing of used compressor lubricating oil.*

The EPA has divided the refrigeration and air conditioning industry into four classifications for licensing purposes. The classifications are Type I (small appliance), Type II (high pressure), Type III (low pressure), and Type IV (universal). In this chapter we discuss low-pressure equipment,

99

high-pressure equipment holding more than 50 pounds of refrigerant, and some high-pressure equipment holding less than 50 pounds of refrigerant.

These types of systems make up the majority of applications. Chillers, both high and low pressure, contain extremely large amounts of refrigerant for proper operation. Large systems that use some type of noncentrifugal compressor also need a large charge of refrigerant, requiring much attention. Most commercial refrigeration installations also use a large refrigerant charge. Current and planned experiments attempt to determine what refrigerant should be used, the type of oil to use, and the types of equipment modifications required to make the system operate with the new refrigerant.

Low-Pressure Chillers

These types of systems use CFC-11. When the decision is made to retrofit a CFC-11 chiller, the choice is generally HFC-123. The CFC-11 charge is removed and a proper charge of the new refrigerant is placed into the system. However, several retrofit procedures usually are necessary.

The type of refrigerant and all its ramifications must be considered. HFC-123 is toxic and must be vented to the outside of the building in case of excessive pressure buildup inside the system. HFC-123 has a threshold limit value (TLV) of 10 parts per million. There will be an efficiency loss of approximately 2% to 5%, and a capacity reduction of about 3%. A full equipment room monitoring system is required in the case of refrigerant leaks. A halogen selective type is suggested. HFC-123 does have very low GWP (global warming potential) and ODP (ozone depletion potential) factors.

When replacing CFC-11 with HFC-123, several equipment changes must be made in order for the equipment to operate at peak efficiency. In addition to the requirements of the equipment manufacturer, other changes are needed because of the type of refrigerant being substituted. The compressor lubricating oil probably will need to be changed—the same type will probably be suitable for use with HFC-123. The system gaskets and "O"-rings must be changed to a material compatible with HFC-123. The hermetic motor winding must be rewound with the proper insulation for HFC-123 or the motor replaced with one that is already properly insulated.

Some or all of the added features discussed under Containment in Chapter 4 should be considered. This will make the system as leak-free as possible.

Refrigerant Recovery

The liquid and vapor refrigerant recovery methods discussed later in this chapter should be followed.

Containers for Recovered Fluorocarbon Refrigerants CFC-11 and CFC-113

Steel drums shall comply with the following: Title 49 CFR, DOT specifications 17C or 17E as set forth in Title 49 CFR 178.115 or 178.116.

Containers that originally contained new refrigerant CFC-11 or refrigerant CFC-113 (excluding those originally used for cleaning agents) may be used, provided certain conditions are met.

- The drums shall be inspected internally and externally and found to be clean and free of dents, bulges, holes, cracks, rust, pits, creases, or other structural weaknesses.
- Closure devices, including gaskets, shall be in such condition that they comply, in all respects, with the original requirements for the drum.
- Drums that originally contained CFC-11 or CFC-113 must be made to comply with Section 6.5.3. Previous labels and markings must be removed and be replaced with new labels and markings per Section 6.3.
- The drum shall be gray. The top head shall be yellow.

101

Recovery equipment purchased after November 15, 1993, and used for refrigerant recovery on low-pressure systems must be able to pump a 29-inch vacuum. However, EPA grandfathered recovery equipment that pumps a 25-inch vacuum and that was in use before that time. Before opening a low-pressure system, the pressure inside the system must be increased to 0 psig. This prevents pulling moisture and air into the system.

High-Pressure Chillers

Most of these types of systems currently use either CFC-12 or CFC-500. These refrigerants are being replaced with either HCFC-22 or HFC-134a. When the decision to retrofit this type of chiller is made, the decision for the type of refrigerant also must be made.

HCFC-22 is under federal control. A federal tax is placed on this refrigerant. HCFC-22 will start being phased out in a few years, requiring another retrofit if this refrigerant is used in a retrofit. Some building owners will not mind this extra expense for several reasons. The retrofit to HCFC-22 costs less now than a retrofit to HFC-134a.

A retrofit to HFC-134a will have a higher initial cost, but the expense occurs one time. HFC-134a is free from future refrigerant restrictions. It has zero ODP and GWP potentials. It is suggested that HFC-134a be vented to the outside in case of excessive pressure buildup inside the system. When changing to HFC-134a, the lubricating oil must be changed to the equipment manufacturer's recommendations. The type of oil used may be either an ester type or a PAG (polyalkylene glycol) type. The system must be flushed at least three times, or until only about 1% or less of the old oil type remains in the system. Common leak detectors do not pinpoint HFC-134a leaks as readily as they will HCFC-22 and CFC-500. Fluorescent dyes are good leak detectors for HFC-134a, except in areas where the leak may be hidden. The hermetic motor insulation will probably not need to be changed with HFC-134a.

When replacing either CFC-12 or CFC-500 with HFC-134a, the impeller must turn faster or the size of the impeller may be increased to achieve the same results. This is required because of the head. Again consult with the equipment manufacturer about such changes. There will probably be a performance drop of about 2% to 7%.

102

Warning: HFC-134a has been shown to be nonflammable at ambient temperatures and atmospheric pressures. However, tests under controlled conditions indicate that, at pressures above atmospheric and with air concentrations greater than 60% by volume, HFC-134a can form combustible mixtures. (In this regard, the product behaves in a similar manner to HCFC-22, which is also combustible at pressures above atmospheric in the presence of high air concentrations. The tests cause the belief that HFC-134a can undergo combustion at lower pressures than HCFC-22.)

Some facilities could conceivably have oxygen cylinders at the site for welding or other purposes. Although data are not available on the combustibility of HFC-134a with oxygen, experience with other materials indicates that combustibility would be enhanced. Note the following:

- Precautions should be taken against inadvertent connection of oxygen lines with HFC-134a process equipment.
- Compressed air should not be connected to HFC-134a process equipment.
- Avoid mixing HFC-134a with air or oxygen.
- Avoid mixing refrigerants.

Refrigerant Removal (Recovery)

When repairs are to be made or a retrofit is planned, the CFC-11 must be removed from the system. After July 1, 1992, it became unlawful to

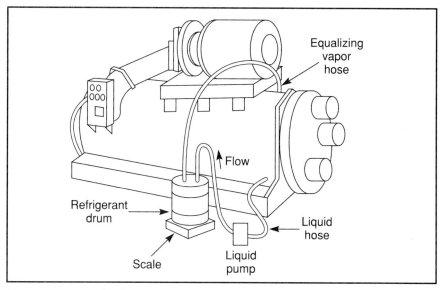

Figure 7-1   Liquid refrigerant pump connections for large amounts of refrigerant.

103

purge the refrigerant into the atmosphere. The fastest and most efficient way to remove the refrigerant is to pump it out in liquid form. Be sure to use a liquid refrigerant pump to prevent damage to the pump Figure 7-1.

Most of the liquid refrigerant is removed first, and then the remaining refrigerant is removed with a conventional recovery unit. If the unit is equipped with a liquid receiver and has the proper service valves, the refrigerant may be pumped into the receiver in liquid form. The liquid may then be pumped out with a liquid refrigerant pump. However, if the compressor is inoperative or the proper service valves are not provided, the refrigerant cannot be removed in this manner.

Vapor Recovery

Connect the vapor recovery unit as shown in Figure 7-2.

The vapor is drawn into the recovery unit, compressed, condensed, and then forced into a refrigerant recovery drum, or cylinder. Notice that there is only one connection to the system. This method is used when only vapor refrigerant is to be removed from the system.

Recovery equipment purchased after November 15, 1993, and used for refrigerant recovery on high-pressure systems must be able to pump a 10-inch to 15-inch vacuum. EPA intends to grandfather recovery equipment that will pump a 4-inch vacuum that is in use before January 1, 1993. However, these provisions require that the equipment be certified before January 1, 1994.

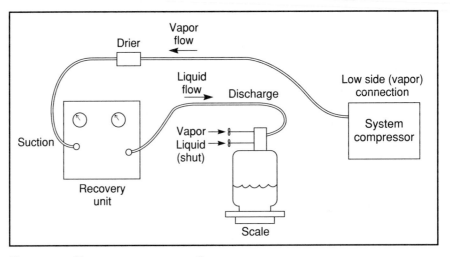

Figure 7-2   Vapor recovery connections.

Specifications for Containers for Recovered Fluorocarbon Refrigerants

104

ARI Guideline K is the guide for cylinders used in recovering CFC refrigerants. The containers, color code, and filling procedure follows. (For complete coverage of these specifications, see the ARI Guideline K.)

Containers

Cylinders for recovered fluorocarbon refrigerants CFC-12, HCFC-22, CFC-114, CFC-500, and CFC-502 shall comply with the following:

United States Department of Transportation (DOT) specification packaging, see Title 49 CFR (Code of Federal Regulations), and have a service pressure rating not less than 325 psig.

Note: Previously filled DOT specification 39 *disposable* cylinders shall not be used for the storage and/or transportation of recovered fluorocarbon refrigerants. Federal law forbids transportation of specification 39 disposable cylinders if refilled. Penalty for violating this requirement is up to $25,000 fine and 5 years' imprisonment [Title 49 U.S.C. (United States Code) Sec. 1809].

The valves shall comply with Compressed Gas Association Standard V-1, "Compressed Gas Cylinder Valve Outlet and Inlet Connections."

Safety Relief devices shall comply with Pressure Relief Standard Part 1—Cylinders for Compressed Gases, Compressed Gas Association Pamphlet S-11.

Cylinder Color Code

Following are examples of coloring schemes for various recovery containers. Depending upon the provider of the recovery container, the actual coloring may vary. However, the use of yellow similar to the following examples will identify the container as a recovery vessel.

Cylinders with Nonremovable Collars. The body shall be gray. The collar shall be yellow.

Cylinders with Removable Caps. The body shall be gray. The shoulder and the cap shall be yellow.

Filling Procedure

Important: Do not mix refrigerants when filling containers.

Cylinders and Ton Tanks. Do not fill if the present date is more than 5 years past the test date on the container. The test date is stamped on the shoulder or collar of cylinders and on the valve end chime of ton tanks and appear as follows:

105

$$A1$$
$$12 \qquad 89$$
$$32$$

This indicates that the cylinder was retested in December of 1989 by retester number A123.

Cylinders and tanks shall be weighed during filling to ensure user safety. The maximum gross weight is indicated on the side of the cylinder and shall never be exceeded.

Cylinders shall be checked for leakage prior to shipment. Leaking cylinders must not be shipped.

In most instances the cylinder is used to recover the refrigerant from several appliances before taking it to the reclamation center.

Leak Testing

Leak testing may be done by introducing dry nitrogen into the system and using a "soap bubble" test to locate any leaks. CAUTION: Do not use oxygen for this procedure; an explosion could occur.

Usually a liquid plastic leak detector is used for this procedure. The solution is mopped around all of the suspected leaks and observed for several minutes. If a leak is present, bubbles will form around the suspected area. If the leak cannot be found in this manner, introduce a small amount of HCFC-22 into the system and use a halide leak detector to locate the leak. Be sure to leak test the complete system so that all leaks can be repaired at the same time without repeatedly recovering the refrigerant. When the leak or leaks are located, recover the refrigerant from the system with the recovery unit. Do not mix this nitrogen/refrigerant mixture with any other refrigerant. Place it in a cylinder especially for this purpose and save it so that it can be used at a later date for leak testing. Be sure to clean the mixture each time it is used to prevent the introduction of moisture and other contaminants into the system.

The system is now evacuated to the recommendations of the equipment manufacturer. The system is then recharged with the proper type and amount of refrigerant.

Noncentrifugal Compressor Systems

106

These types of systems range in size from about $1^1/_2$ tons to several hundred tons of capacity. They can be used for commercial air conditioning, commercial refrigeration, residential air conditioning, or for industrial applications. The larger sizes of these systems contain quite large amounts of refrigerant. The refrigerant type may be one of several types of CFCs or HCFCs. In either case, after July 1, 1992, it became unlawful to vent the refrigerant to the atmosphere. Therefore, when any repair or a retrofit is required that does not allow the refrigerant to be contained in the system, the refrigerant must be recovered.

Systems With Service Connections

The larger sizes of these units have service valves to aid in servicing and repair procedures. The smaller systems may have Schrader valves or no service valves. When service valves are provided, it is a relatively simple procedure to connect the recovery unit to the system. When Schrader valves slow the process due to the restriction caused by the valve core, it may be possible to remove the valve core without loss of refrigerant by using a special adaptor for this purpose. The recovery process will then go faster.

Systems Without Service Connections

When smaller systems having no service connections are to be serviced, several methods can be used. A solder-on line tap valve or a bolt-on line tap valve can be used. The solder-on type can be left on after the service procedures are completed. The bolt-on type is subject to leaks, and it is not recommended that they be left on the system.

A Schrader valve can be installed in the system after the refrigerant has been recovered. The gauge hose is connected to the line tap valve. The piercing screw is then screwed in, piercing the line and allowing the refrigerant to enter the gauge hose. The system can then be serviced in the regular manner.

Note: Do not overheat the refrigerant line when installing these valves for two reasons. First, the line may become hot enough to allow the refrigerant pressure inside to blow out the overheated part of the refrigerant line, causing personal injury and possible property damage. Second, the overheated line may cause the refrigerant inside to overheat causing a refrigerant breakdown (the formation of hydrochloric acid).

Another method is to solder a Schrader valve into the process tube on the compressor or on the suction line. However, to use this method the system must be completely out of refrigerant.

When access to the refrigeration system has been accomplished, the refrigerant can be removed, repairs or modifications made, and the system evacuated and recharged with refrigerant.

107

Liquid Recovery

When a large amount of liquid refrigerant is in the system, the most efficient method of recovery is to remove the liquid first, then recover the vapor. To remove the liquid from one of these types of systems, use a recovery/recycle unit that is capable of removing the liquid. To connect the recovery/recycle unit to the equipment, see Figure 7-3.

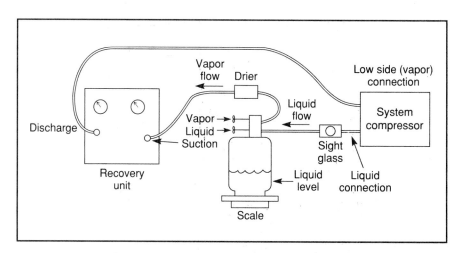

Figure 7-3 Liquid recovery connections.

Make certain that the liquid refrigerant does not enter the compressor of the recovery/recycle unit causing damage to it. In this method the liquid refrigerant flows into the cylinder because of the pressure difference. The suction of the recovery/recycle unit lowers the pressure in the refrigerant recovery cylinder and the discharge from the recovery/recycle unit causes a higher pressure on the liquid inside the system causing it to flow into the cylinder. Notice the drier in the vapor line to the recovery/recycle unit. This will aid in cleaning the refrigerant.

Refrigerant recovery equipment purchased after November 15, 1993, and to be used on high-pressure systems should pump a vacuum of between 10 and 15 inches, depending on the size of the unit being serviced. However, EPA grandfathered recovery/recycling equipment in use before that time if the unit pumps a vacuum of at least 4 inches for high-pressure systems. However, these provisions require that the equipment be certified before January 1, 1994.

When a very large amount of liquid refrigerant is to be recovered, a liquid pump will speed up the recovery process. Connect the liquid pump to the system at a point where the liquid will accumulate, such as a receiver. After the liquid has been removed the vapor can then be removed with a conventional recovery/recycle unit.

108

Vapor Recovery

To recover the refrigerant vapor from a system to be repaired connect the recovery unit as shown in Figure 7-2. The vapor is drawn into the recovery unit, compressed, condensed, and then forced into the cylinder. Notice that there is only one connection to the system. This method is suited for a smaller system where there is very little liquid refrigerant to be removed. Notice the drier in the vapor line to the recovery unit. This will aid in cleaning the refrigerant.

Specifications for Containers for Recovered Fluorocarbon Refrigerants

For these specifications and others concerning refrigerant cylinders see this heading presented earlier in this chapter.

Leak Testing

After the repairs or modifications have been made, the system is ready for leak testing. This can be accomplished by introducing dry nitrogen into the system until the near operating pressure is reached for the type of refrigerant used in the system. All suspected areas are then wet with "soap bubbles"—a liquid plastic leak detector. The suspected areas are then observed for several minutes for bubbles. If none appear the area is not leaking. However, if the system loses pressure after setting idle for a couple of hours there is a leak that must be found.

If a leak is still present, introduce a small amount of HCFC-22 refrigerant into the system. Allow the unit to stand idle long enough for the two gases to mix. Then leak test with a halide leak detector. Be sure to locate all leaks before removing the refrigerant/nitrogen mixture to save time and expense. Recover the refrigerant/nitrogen mixture into a cylinder specifically for this purpose. Do not mix refrigerants or mix this mixture with other refrigerants. Save the mixture for future leak testing.

After all repairs or modifications have been made, evacuate the system and recharge it with the proper refrigerant.

Refrigerant Replacements

Usually the replacement refrigerant for CFC-500 and CFC-502 is either HCFC-22 or HFC-134a. When HCFC-22 is used the modifications are not as extensive as when HFC-134a is used. Consult with the equipment manufacturer for recommendations concerning refrigerant replacement.

109

HCFC-22 Replacement. When HCFC-22 is the replacement refrigerant, the refrigerant controls usually need to be either replaced or adjusted for the new pressures encountered. When changing to HCFC-22, this may be a short-lived solution, because of its phaseout date.

HFC-134a. When HFC-134a is the replacement refrigerant, modification probably will be needed. Usually the gaskets, "O"-rings, and oil must be changed. Changing the gaskets can be quite an extensive process. It may be less expensive to replace the equipment, especially if it is old and not in perfect condition.

Changing the Compressor Oil. The lubricating oil must be replaced when HFC-134a is the replacement refrigerant. The oil is changed to either an ester or a PAG type that is compatible with HFC-134a. The oil must be changed at least three times, to remove enough of the old oil to reach an acceptable level for system performance, usually less than 1%. This can be rather expensive, but it may be worth the trouble if the system is in excellent condition.

The oil that is removed from the system must be disposed of properly. If it is not too contaminated it can be refined to meet the required specifications. In any case all local, state, and federal regulations must be followed. The federal regulations are listed in Title 40 CFR 266 under subpart H—Hazardous Waste Burned in Boilers and Industrial Furnaces: Section 266.107 Standards to Control Hydrogen Chloride (HCl) and Chlorine Gas (Cl$_2$) Emissions.

Summary

- The four categories of licensing for refrigerant service by EPA are low-pressure equipment, high-pressure equipment containing more than 50 pounds of refrigerant, high-pressure equipment containing less than 50 pounds of refrigerant, and small appliances.

- Low-pressure chillers currently use CFC-11 as the refrigerant. When the decision is made to retrofit a CFC-11 chiller, the choice is generally to use HFC-123. The CFC-11 charge is removed and a proper charge of new refrigerant is placed into the system. Retrofit procedures usually are necessary.

- HFC-123 is toxic and must be vented to the outside of the building in case of excessive pressure buildup inside the system.

- HFC-123 has a threshold limit value (TLV) of 10 parts per million.

- There will be an efficiency loss of approximately 2% to 5%, and a capacity reduction of about 3% when a chiller is changed to HFC-123.

- Full equipment room monitoring system is required in the case of refrigerant leaks. A halogen selective type is recommended. HFC-123 has low GWP and ODP factors.

- When replacing CFC-11 with HFC-123, several equipment changes must be made for the equipment to operate at peak efficiency.

- The compressor lubricating oil will probably need to be changed— the same type will probably be suitable for use with HFC-123.

- The system gaskets, or "O"-rings, must be of a material compatible with HFC-123.

- The hermetic motor winding must be rewound with wire having the proper insulation for HFC-123 or the motor replaced with one that is already properly insulated.

- Containers that originally contained refrigerant CFC-11 or refrigerant CFC-113 (excluding those originally used for cleaning agents) may be used. Previous labels and markings must be removed and replaced with new labels and markings.

- HFC-123 recovery cylinders are painted with a gray drum and a yellow head.

- Recovery/recycle equipment purchased after January 1, 1993, and to be used for refrigerant recovery on low-pressure systems must be capable of pumping a 29-inch vacuum.

- Most high-pressure chillers use either CFC-12 or CFC-500. These refrigerants are being replaced with either HCFC-22 or HFC-134a. When the decision to retrofit this type of equipment is made, the decision for the type of refrigerant also must be made.

110

◆ HCFC-22 is also under federal control at this time. A federal tax is placed on this refrigerant, which is being phased out.

◆ It is suggested that HFC-134a be vented to the outside in case of excessive pressure buildup inside the system. When changing to HFC-134a, the compressor lubricating oil must be changed to the equipment manufacturer's recommendations. The type of oil may be either an ester or PAG type. When changing types of oil, the system must be flushed at least three times, or until only about 1% or less of the old type oil remains in the system.

◆ Common leak detectors will not pinpoint HFC-134a leaks as readily as they will HCFC-22 and CFC-500. Fluorescent dyes make good leak detectors for HFC-134a, except in areas where the leak may be hidden.

◆ HFC-134a has been shown to be nonflammable at ambient temperatures and atmospheric pressure. Recent tests under controlled conditions indicate that, at pressures above atmospheric and with air concentrations greater than 60% by volume, HFC-134a can form combustible mixtures.

◆ Recovery equipment purchased after January 1, 1993, and used for refrigerant recovery on high-pressure equipment must be able to pump a 10- to 20-inch vacuum.

111

◆ Previously filled DOT 39 disposable cylinders shall not be used for storage and/or transportation of recovered fluorocarbon refrigerants.

◆ System leak testing may be done by introducing dry nitrogen into the system to near operating pressures and then using the soap bubble test to locate any leaks.

◆ If the leak cannot be found in this manner, introduce a small amount of HCFC-22 into the system and allow the system to stand idle enough time for the two gases to mix. Then use a halide leak detector to locate the leak. Be sure to leak test the complete system so that all leaks can be repaired at the same time without repeatedly recovering the nitrogen/refrigerant mixture. When the leaks have all been found, recover the nitrogen/HCFC-22 from the system with the recovery unit. Do not mix this nitrogen/refrigerant mixture with any other refrigerant. Place it in a cylinder especially for this purpose and save it so that it can be used at a later date for leak testing. Be sure to clean the mixture each time it is used to prevent the introduction of moisture and other contaminants into the system.

◆ After July 1, 1992, it became unlawful to vent refrigerant to the atmosphere. Therefore, when any repair or retrofit is required that will not allow the refrigerant to be contained in the system, the refrigerant must be recovered.

◆ Systems having no service connections can be serviced, using a solder-on type line-tap valve.

- ◆ When recovering liquid refrigerant with a recovery/recycle unit, make certain that the liquid does not enter the compressor of the unit, causing damage to the compressor.

- ◆ Refrigerant recovery equipment purchased after January 1, 1993, and to be used on high-pressure systems should be capable of pumping a vacuum between 10 inches and 20 inches depending on the size of the unit being serviced. When recovering refrigerant vapor, the vapor is drawn into the recovery unit, compressed, condensed, and then forced into the cylinder. This method is suited for smaller systems when there is very little liquid refrigerant to be removed.

◆ Key Terms ◆

ARI (Air Conditioning and Refrigeration Institute)

CFR (Code of Federal Regulations)

DOT (Department of Transportation)

EPA (Environmental Protection Agency)

GWP (global warming potential)

ODP (ozone depletion potential)

PAG (polyalkylene glycol)

Review Questions

Essay

1. What type of refrigerant is usually used to replace CFC-11 in low-pressure chillers?

2. What is the approximate efficiency loss when changing from CFC-11 to HFC-123?

3. What is required in the equipment room when HFC-123 refrigerant is used?

4. What type of refrigerant containers are used for HFC-123?

5. When HFC-123 is used as the replacement refrigerant for CFC-11 should the oil also be changed?

6. Will HFC-123 attack the motor winding insulation on hermetic motors?

7. What is the color of HFC-123 refrigerant recovery cylinders?

8. What type of vacuum will recovery equipment purchased after January 1, 1993, and to be used for refrigerant recovery on low-pressure systems, have to pump?

9. In high-pressure equipment, with what are refrigerants CFC-12 and CFC-500 being replaced?

10. After July 1, 1992, is it lawful to vent a partial refrigerant charge to the atmosphere?

Fill-in-the-Blank

11. _____ is also under federal control at this time.

12. _____ refrigerant must be vented to the outside of the building.

13. It is suggested that _____ be vented to the outside of the building.

14. When a system is retrofitted with HFC-134a the _____ _____ must also be changed.

15. Common leak detectors will not readily pinpoint _____ leaks.

True-False

16. Previously filled DOT 39 disposable cylinders shall not be used for storage and/or transportation of fluorocarbon refrigerants.

17. When Schrader valves are used on the system, the recovery process is slowed down because of their restriction.

18. Refrigerant recovery equipment purchased after January 1, 1993, and to be used on high-pressure equipment, shall pump a vacuum between 10 and 20 inches Hg.

19. There are no requirements for the disposal of used refrigeration oil.

20. HFC-123 has a low toxic rating.

113

Multiple Choice

21. When recovering liquid refrigerant with a recovery/recycle unit make certain that the liquid
    a. does not enter the recovery/recycle unit compressor.
    b. has no moisture in it.
    c. is contaminated with noncondensables.
    d. does not contain refrigeration oil.

22. The refrigerant HFC-123 has a threshold limit value (TLV) of
    a. 0 ppm.
    b. 5 ppm.
    c. 10 ppm.
    d. 15 ppm.

23. The type of oil used when retrofitting a system with HFC-134a is
    a. mineral oil.
    b. ester or PAG.
    c. not changed.
    d. more susceptible to moisture contamination.

24. The penalty for storage and/or transporting previously filled DOT 39 cylinders is
    a. very small.
    b. $10,000 and 5 years imprisonment.
    c. $25,000 and 1 year imprisonment.
    d. $25,000 and 5 years imprisonment.

25. When retrofitting a system to HFC-134a, the oil should be changed
    a. one time.
    b. frequently until only about 1% or less of the old oil remains in the system.
    c. frequently until about 10% or less of the old oil remains in the system.
    d. two times.

114

Residential Refrigeration and Air Conditioning

Objectives

After completion of this chapter you should:

♦ *know more about the licensing requirements.*

♦ *be more familiar with the requirements of Section 608 of the National Recycling and Emission Reduction Program and how it affects domestic appliances.*

♦ *know more about the safe disposal of appliances containing Class I and Class II substances.*

♦ *know more about the prohibitions placed on Class I and Class II substances.*

♦ *know more about the refrigerant recovery procedures used in domestic appliance service and repairs.*

The EPA divides the refrigeration and air conditioning industry into four classifications for licensing requirements: Type I (small appliance), Type II (high pressure), Type III (low pressure), and Type IV (universal). In this chapter we will discuss the high-pressure equipment containing less than 50 pounds and small appliance recovery/recycle procedures.

Domestic refrigerators, freezers, and window air conditioning units are also covered by the National Recycling and Emission Reduction Program. Even the small amount of refrigerant contained in these sys-

115

tems must be recovered and recycled with the same requirements that are placed on the larger systems.

Section 608 of the National Recycling and Emission Reduction Program in part states the following:

(a) IN GENERAL: (1) The Administrator shall, by not later than January 1, 1992, promulgate regulations establishing standards and requirements regarding the use and disposal of appliances and industrial process refrigeration. Such standards and requirements shall become effective not later than July 1, 1992.

(2) The Administrator shall, within four years after the enactment of the Clean Air Act Amendments of 1990, promulgate regulations establishing standards and requirements regarding use and disposal of Class I and Class II substances not covered by paragraph (1), including the use and disposal of Class II substances during service, repair, or disposal of appliances and industrial process refrigeration. Such standards and requirements shall become effective not later than 12 months after promulgation of the regulations.

(3) The regulations under this subsection shall include requirements that:

(A) reduce the use and emission of such substances to the lowest achievable level (LAL), and

(B) maximize the recapture and recycling of such substances. Such regulations may include requirements to use alternative substances (including substances that are not Class I or Class II substances) or to minimize use of Class I or Class II substances, or to promote the use of safe alternatives pursuant to Section 612 or any combination of the foregoing.

(b) SAFE DISPOSAL: The regulations under subsection (a) shall establish standards and requirements for the safe disposal of Class I and Class II substances. Such regulations shall include each of the following:

(1) Requirements that Class I or Class II substances contained in bulk in appliances, machines, or other goods shall be removed from each such appliance, machine, or other good prior to the disposal of such items or their delivery for recycling.

(2) Requirements that any appliance, machine, or other good containing a Class I or Class II substance in bulk shall not be manufactured, sold, or distributed in interstate commerce unless it is equipped with a servicing aperture or an equally effective design feature which will facilitate the recapture of such substance during service and repair or disposal of such items.

(The full section is printed in Chapter 3 of this text.)

(3) Requirements that any product in which a Class I or Class II substance is incorporated so as to constitute an inherent element of such product shall be disposed of in a manner that reduces, to the

116

maximum extent practicable, the release of such substance into the environment. If the Administrator determines that the application of the paragraph to any product would result in producing only insignificant environmental benefits, the Administrator shall include in such regulations an exception for such product.

(c) PROHIBITIONS: (1) Effective July 1, 1992, it shall be unlawful for any person, in the course of maintaining, servicing, repairing, or disposing of an appliance or industrial process refrigeration, to knowingly vent or otherwise knowingly release or dispose of any Class I or Class II substance used as a refrigerant in such appliance (or industrial process refrigeration) in a manner which permits such substance to enter the environment. De minimis releases associated with good faith attempts to recapture and recycle or safely dispose of any such substance shall not be subject to the prohibition set forth in the preceding sentence. The fine for intentional venting of refrigerant beginning July 1, 1992 is up to $25,000 per day per offense. Each kilogram is considered to be a separate offense. Anyone can report a violation. If the person venting the refrigerant is prosecuted and fined, the one who reported the violation can receive up to $10,000.

(2) Effective five years after the enactment of the Clean Air Act Amendments of 1990, paragraph (1) shall also apply to the venting, release, or disposal of any substance for a Class I or a Class II substance by any person maintaining, servicing, repairing, or disposing of an appliance or industrial process refrigeration unit which contains and uses as a refrigerant any such substance, unless the Administrator determines that venting, releasing, or disposing of such substance does not pose a threat to the environment. For purposes of this paragraph, the term "appliance" includes any device which contains and uses as a refrigerant a substitute substance and which is used for household or commercial purposes, including any air conditioner, refrigerator, chiller, or freezer.

117

Methods of Refrigerant Recovery

The two procedures used to recover refrigerant from domestic refrigerators, freezers, and window air conditioning units are recovery into a cylinder and recovery into a plastic bag.

Refrigerant Recovery Procedure

In the service of domestic refrigerators, freezers, and window air conditioners, there are procedures one must use to prevent the venting of

CFCs to the atmosphere. The following discussion explains these procedures.

Refrigerators and Freezers

These appliances will be treated at the same time because they are so related. Most of these appliances manufactured to present date lack refrigerant service valves. Therefore, the service technician must install one before attempting to alter the refrigerant charge in these appliances. Federal regulations prohibit future manufacturing of these appliances without service fittings for the refrigeration system. Usually, a service technician will install a valve only on the suction line to the compressor. However, there may be times when a valve will be required on both the suction and discharge lines. The same procedure is used on both lines. There are several types of valves on the market, such as the solder-on type and the bolt-on type line tap piercing valves. The solder-on type is preferred over the bolt-on type because the solder-on type is leak-free when properly installed, Figure 8-1.

118

Note: Do not overheat the refrigerant line when installing these valves for two reasons: First the line may become hot enough to allow the refrigerant pressure inside to blow-out the overheated refrigerant line, causing personal injury and possible property damage. Second, the overheated line may cause the refrigerant inside to overheat causing a refrigerant breakdown (the formation of hydrochloric acid). If a bolt-on line tap valve is used, a Schrader valve can be installed after the refrigerant has been recovered from the unit. It would be a good practice to remove the bolt-on type and seal the puncture hole to prevent as many future leaks as possible.

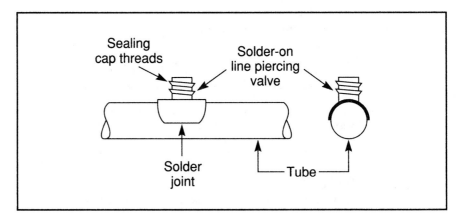

Figure 8-1   Solder-on line piercing valves.

After the line piercing valve has been installed, the required service to the refrigerant system can then be performed without venting refrigerant. The gauge manifold hose is now connected to the line piercing valve and the line is pierced by screwing the piercing screw into the refrigerant line. When the line has been pierced, a pressure is indicated on the refrigeration gauge. It is usually best to back out the screw a little way so that the full pressure can be indicated on the gauges and also so that the service procedure will progress much faster.

The recovery unit can now be connected to the system. Since there will, in most cases, be only a small amount of refrigerant in these systems, very little, if any, liquid refrigerant will remain in the system after the refrigerant pressure has equalized. Therefore, only the vapor removal procedure will be discussed.

Vapor Recovery

To recover the refrigerant vapor from a system to be repaired, connect the recovery/recycle unit as shown in Figure 8-2.

The vapor is drawn into the recovery/recycle unit, compressed, condensed, and then forced into the recovery cylinder. Notice that there is only one connection to the system. This method is suited for use when only vapor or a very small amount of liquid refrigerant is to be recovered. Notice the drier in the vapor line to the recovery/recycle unit. This will aid in cleaning the refrigerant.

119

The recovery/recycle equipment purchased for use on these types of systems after November 15, 1993, must pump a vacuum of approxi-

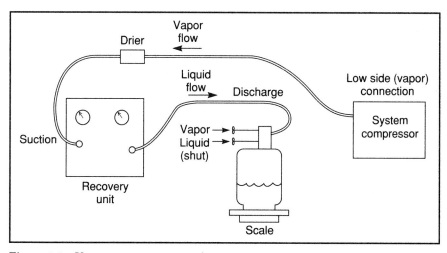

Figure 8-2   Vapor recovery connections.

mately 10 inches and must be able to remove 80% to 90% of the refrigerant in the system. However, the EPA intends to grandfather systems that were in use before that date if they can pump a vacuum of 4 inches or more. However, these provisions require that the equipment be certified before January 1, 1994.

At this point the necessary work can be completed. The system can be evacuated and recharged with the proper amount and type of refrigerant and placed back into operation. The refrigerant should be recycled before charging it back into the system. Some manufacturers of this type of equipment will not honor the warranty if recycled refrigerant is used in their system. Therefore, the manufacturer of the refrigerator or freezer should be contacted for their views on recycled refrigerant.

A point to remember is that the recovered refrigerant and oil are the property of the equipment owner. The owner must decide what is to be done with them. If the refrigerant is to be used in another unit belonging to the same owner, it can be recycled and saved, or charged into another unit without being reclaimed to standard purity levels. The oil that is removed from the system must be disposed of properly. If it is not too contaminated it can be refined to meet the required specifications. In any case all local, state, and federal regulations must be followed. The federal regulation is listed in Title 40 CFR 266 under subpart H-"Hazardous Waste Burned in Boilers and Industrial Furnaces," Section 266.107 "Standards to Control Hydrogen Chloride (HCl) and Chlorine Gas (Cl_2) Emissions."

Recovery into a Bag

This method recovers the refrigerant into a special plastic bag rather than into a cylinder. It is an expensive procedure due to the cost of the bag for each unit, and each bag must be handled individually. The refrigeration system is connected to the bag and the pressure is allowed to equalize into the bag. This procedure will recover approximately 80% to 90% of the refrigerant. The remainder of the refrigerant can then be recovered with a recovery unit.

There can be no liquid refrigerant in the bag after the recovery process. When filled the bag is shipped by special handling. Be sure to follow the procedures that accompany the bag before attempting to recover the refrigerant with this method.

Specifications for Containers for Recovered Fluorocarbon Refrigerants

ARI Guideline K is the guide for cylinders used in recovering CFC and HCFC refrigerants. The containers, color code, and filling procedure follows. (For a complete coverage of these specifications see the ARI Guideline K.)

Containers

Cylinders for Recovered Fluorocarbon Refrigerants CFC-12, HCFC-22, CFC-114, CFC-500, and CFC-502 shall comply with the following:

United States Department of Transportation (DOT) specification packaging, see Title 49 CFR (Code of Federal Regulations), and have a service pressure rating not less than 325 psig.

Previously filled DOT specification 39 *disposable* cylinders shall not be used for the storage and/or transportation of recovered fluorocarbon refrigerants. Federal law forbids transportation of specification 39 disposable cylinders if refilled. Penalty for violating this requirement is up to $25,000 fine and 5 years' imprisonment [(Title 49 U.S.C. (United States Code) Sec. 1809].

The valves shall comply with Compressed Gas Association Standard V-1, "Compressed Gas Cylinder Valve Outlet and Inlet Connections."

Safety relief devices shall comply with "Pressure Relief Standard Part 1—Cylinders for Compressed Gases," Compressed Gas Association Pamphlet S-11.

121

Cylinder Color Code

Following are examples of coloring schemes for various recovery containers. Depending upon the provider of the recovery container, the actual coloring may vary. However, the use of yellow similar to the following examples will identify the container as a recovery vessel.

Cylinders with Nonremovable Collars: The body shall be gray. The collar shall be yellow.

Cylinders with Removable Caps: The body shall be gray. The shoulder and the cap shall be yellow.

Filling Procedure

IMPORTANT: DO NOT MIX REFRIGERANTS WHEN FILLING CONTAINERS.

Cylinders: Do not fill if the present date is more than five years past the test date on the container. The test date will be stamped on the shoulder or collar of cylinders and appear as follows:

A1

12   89

32

This indicates that the cylinder was retested in December of 1989 by retester number A123.

Cylinders shall be weighed during filling to ensure user safety. The maximum gross weight is indicated on the side of the cylinder and shall never be exceeded.

Cylinders shall be checked for leakage prior to shipment. Leaking cylinders must not be shipped.

In most instances the cylinder is used to recover the refrigerant from several appliances before taking it to the reclamation center.

Leak Testing

Leak testing may be done by introducing dry nitrogen into the system to increase the pressure to near operating pressure and using a "soap bubble" test to locate any leaks. CAUTION: Do not use oxygen for this procedure because an explosion could occur.

Usually a liquid plastic leak detector is used for this procedure. The solution is mopped around all of the suspected leaks and observed for several minutes. If a leak is present, bubbles will form around the suspected area. If the leak cannot be found in this manner, introduce a small amount of HCFC-22 into the system and allow it to stand idle long enough for the two gases to mix. Then use a halide leak detector to locate the leak. Be sure to leak test the complete system so that all leaks can be repaired at the same time without repeatedly recovering the refrigerant. When the leak, or leaks, are located, recover the refrigerant from the system with the recovery unit. Do not mix this nitrogen/refrigerant mixture with any other refrigerant. Place it in another cylinder especially for this purpose and save it so that it can be used at a later date for leak testing. Be sure to clean the mixture each time it is used to prevent the introduction of moisture and other contaminants into the system.

The system is now evacuated to the recommendations of the equipment manufacturer. The system is recharged with the proper type and amount of refrigerant.

HFC-134a Refrigerant

If the refrigerant type is to be changed, it will probably be replaced with HFC-134a. There are some considerations that require attention before attempting to recharge the system with HFC-134a. The compressor lubricating oil must be changed and the system monitored until all of the old oil is taken from the system, or at least removed down to 1% or less remaining in the system. This usually requires three oil changes. This is quite an expensive changeover. It is usually less expensive to replace the equipment, unless it is some type of specialized application. All of this must be taken into account before changing the refrigerant type. HFC-134a is not a drop-in replacement refrigerant for CFC-12.

WARNING: HFC-134a has been shown to be nonflammable at ambient temperature and atmospheric pressure. However, recent tests under controlled conditions have indicated that at pressures above atmospheric and with air concentrations greater than 60% by volume, HFC-134a can form combustible mixtures. (In this regard, the product behaves in a similar manner to HCFC-22, which is also combustible at pressures above atmospheric in the presence of high air concentrations. The tests cause the belief that HFC-134a can undergo combustion at lower pressures than HCFC-22.)

Some facilities could conceivably have oxygen cylinders at the site for welding, etc. Although data are not available on the combustibility of HFC-134a with oxygen, experience with other materials indicates that combustibility would be enhanced. Note the following:

- Precautions should be taken against inadvertent connection of oxygen lines with HFC-134a process equipment.
- Compressed air should not be connected to HFC-134a process equipment.
- Avoid mixing of HFC-134a with air or oxygen.
- Avoid mixing of refrigerants.

123

Window Air Conditioning Units

Window air conditioning units are self-contained units that are placed in the window or other opening, and are used to cool only one or two rooms as compared to central air conditioning units that are used to cool complete buildings. These units generally use HCFC-22 as the refrigerant, therefore changing the refrigerant is not a concern. However, the EPA ruled that it shall not be vented to the atmosphere after July 1, 1992.

The window air conditioning units that have been previously manufactured are not equipped with refrigerant service valves. Therefore, some type of service valve must be installed by the service technician. Usually, a line tap valve is the best to install. These valves are placed on the refrigerant line and a needle is screwed into the refrigerant line piercing it. There are two types of these valves: the solder-on type and the bolt-on type.

The solder-on type is the best because when properly installed, it does not leak. The bolt-on type can possibly leak at some future date.

When installing the solder-on type use caution to not overheat the refrigerant line. If the line is overheated, the heated portion could possibly blow-out causing personal injury and possible property damage (refer to Figure 8-1). Sometimes, it is easier to install a line piercing valve on the compressor process tube rather than the suction line. The installation procedure is the same. If a service connection is required on the high side, a connection must be installed there also.

Refrigerant Recovery

These units use a relatively small refrigerant charge. Therefore, the liquid recovery method is seldom used. It is simpler and less costly to use the vapor recovery method. Refer to the earlier sections on refrigerant vapor recovery procedures, leak testing procedures, and recharging procedures. Since HCFC-22 will be available for several years, the refrigerant type replacement procedure is not considered here.

124

Specifications for Containers for Recovered Fluorocarbon Refrigerants

See the earlier sections for this and other procedures to be used for refrigerant recovery cylinders.

Summary

◆ The four classifications for servicing refrigeration and air conditioning units are low-pressure equipment, high-pressure equipment containing more than 50 pounds of refrigerant, high-pressure systems containing less than 50 pounds of refrigerant, and small appliances.

◆ Domestic refrigerators, freezers, and window air conditioning units are also covered by the National Recycling and Emission Reduction Program. Even the small amount of refrigerant contained in these systems must be recovered and recycled with the same requirements that are placed on the larger systems.

◆ The regulations shall establish standards and requirements for the safe disposal of Class I and Class II substances. Such substances shall include each of the following.

    1. Requirements that Class I and Class II substances contained in bulk appliances, machines, or other goods shall be removed from each appliance, machine, or other good prior to the disposal of such items or their delivery for recycling.

2. Requirements that any appliance, machine, or other good containing Class I or Class II substance in bulk shall not be manufactured, sold, or distributed in interstate commerce unless it is equipped with a servicing aperture or an equally effective design feature that facilitates the recapture of such substance during service and repair or disposal of such items.

3. Requirements that any product in which Class I or Class II substance is incorporated so as to constitute an inherent element of such a product shall be disposed of in a manner that reduces, to the maximum extent practicable, the release of such substance into the environment.

◆ Effective July 1, 1992, it became unlawful for any person, in the course of maintaining, servicing, repairing, or disposing of an appliance or industrial process refrigeration, to knowingly vent or otherwise knowingly release or dispose of any Class I or Class II substance used as refrigerant in such appliance, or industrial process refrigeration in a manner which permits such substance to enter the environment. De minimis releases associated with good faith attempts to recapture and recycle or safely dispose of any such substance shall not be subject to the prohibition set forth in the preceding sentence.

125

◆ Most domestic refrigerators, freezers, and window air conditioning units previously manufactured are not equipped with refrigerant service valves. Therefore, the service technician must install one before attempting to alter the refrigerant charge in these appliances.

◆ Do not overheat the refrigerant line when installing solder-on tap-a-line valves for two reasons: First the line may become hot enough to allow the refrigerant pressure inside to blow-out the overheated refrigerant line causing personal injury and possible property damage. Second, the overheated line may overheat the refrigerant inside causing a refrigerant breakdown (the formation of hydrochloric acid).

◆ Refrigerant from refrigerators and freezers may be recovered into a cylinder or into a special plastic bag.

◆ To recover the refrigerant vapor from a system to be repaired, properly connect the recovery/recycle unit to the system. The vapor is drawn into the recovery/recycle unit, compressed, condensed, and then forced into the recovery cylinder. This method is best suited for use when only vapor or a very small amount of liquid refrigerant is to be recovered. A drier in the vapor line to the recovery/recycle unit will aid in cleaning the refrigerant.

◆ The recovery/recycle unit purchased for use on these types of systems after January 1, 1993, must pump a vacuum of approximately 10 inches to 20 inches of vacuum depending on the size of the equipment being serviced. They must be able to remove 80% to 90% of the refrigerant from the system.

- When repairs are completed, the system can be evacuated and recharged with the proper type and amount of refrigerant and placed back into operation. The refrigerant should be recycled before charging it back into the system.

- Another method of refrigerant recovery is to recover the refrigerant into a special plastic bag. This method is expensive due to the cost of the bag for each unit, and each bag must be handled individually. This procedure recovers approximately 80% to 90% of the refrigerant; the remainder can then be recovered with a recovery unit.

- Previously filled DOT Specification 39 disposable cylinders shall not be used for storage and/or transportation of recovered fluorocarbon refrigerants.

- Do not mix refrigerants when filling containers.

- Cylinders shall be weighed during filling to ensure user safety. The MAXIMUM GROSS WEIGHT is indicated on the side of the cylinder and it shall never be exceeded.

- Cylinders shall be checked for leaks prior to shipping. Leaking cylinders must not be shipped.

126

- In most instances the cylinder is used to recover refrigerant from several appliances before taking it to the reclamation center.

- Leak testing is done by introducing dry nitrogen into the system to increase the pressure to near operating pressure and using a soap bubble test to locate any leaks. Usually a liquid plastic leak detector is used for this purpose. If the leak cannot be found in this manner, introduce a small amount of HCFC-22 into the system. Allow the system to stand idle for a while to allow the two gases to mix. Then use a halide leak detector to find the leak.

- Before attempting to recharge the system with HFC-134a, the compressor lubrication oil must be changed and the system monitored until all of the old oil is taken from the system, or at least down to approximately 1% remaining in the system. This usually requires three oil changes.

- HFC-134a has been shown to be nonflammable at ambient temperature and at atmospheric pressure. However, recent studies under controlled conditions have indicated that at pressures above atmospheric and with air concentrations greater than 60% by volume, HFC-134a can form combustible mixtures.

- Precautions should be taken against inadvertent connection of oxygen lines with HFC-134a process equipment.

- Compressed air should not be connected to HFC-134a process equipment.

- Avoid mixing of HFC-134a with air or oxygen.

◆ Key Terms ◆

LAL (lowest achievable level)

Class I substance

Class II substance

DOT (Department of Transportation)

Review Questions

Essay

1. Name the four categories of refrigerant handling license.

2. With what device must appliances containing a Class I or a Class II substance be manufactured?

3. Name the two things that could happen if a refrigerant line containing a refrigerant was overheated.

4. What are the pumping requirements for recovery/recycle equipment purchased after January 1, 1993?

5. What is the penalty for violating Section 608 of the National Recycling and Emission Reduction Program?

127

Fill-in-the Blank

6. The EPA regulations establish standards and requirements for the _____ _____ of Class I and Class II substances.

7. Do not _____ the refrigerant lines when installing a tap-a-line valve on a charged unit.

8. The two procedures used to recover refrigerant from a domestic refrigerator or freezer are recover into a _____ and recover into a _____.

9. HFC-134a at pressures above atmospheric and with air concentrations greater than _____ by volume can form a combustible mixture.

10. Avoid mixing of HFC-134a with _____ or _____.

True-False

11. When changing from a CFC refrigerant to HFC-134a, the oil should be changed until the old oil is diluted to approximately 1%.

12. Refrigerant recovery cylinders have a yellow top and the body is painted the color of the new refrigerant.

13. The refrigerant vapor to be recovered is drawn from the system into the recovery/recycle unit, compressed, condensed, and then forced into the receiver tank.

14. Recovery/recycle equipment purchased after January 1, 1993 for use on domestic equipment must pump at least a 20" vacuum.

15. Recovery/recycle units purchased after January 1, 1993 for use on domestic equipment must be able to remove 80% to 90% of the refrigerant in the system.

Multiple Choice

16. There can be no intentional venting of refrigerant after
    a. January 1, 1992.
    b. July 1, 1992.
    c. January 1, 1993.
    d. July 1, 1993.

17. The penalty for storing and/or transporting previously filled DOT 39 disposable cylinders is
    a. $10,000.
    b. $10,000 and 5 years imprisonment.
    c. $25,000.
    d. $25,000 and 5 years imprisonment.

18. Recovery/recycle equipment purchased after January 1, 1993 and to be used on domestic refrigeration equipment must be capable of pumping
    a. 10" to 20" vacuum.
    b. 15" to 25" vacuum.
    c. liquid refrigerant.
    d. in high ambient temperatures.

19. Overheating a refrigerant line with refrigerant inside could cause
    a. hydrochloric acid.
    b. hydrofluoric acid.
    c. sulfuric acid.
    d. the refrigerant to migrate.

20. When refrigerant types are mixed, the refrigerant is usually
    a. refined.
    b. passed through a drier.
    c. destroyed.
    d. used in another unit.

PART III

Refrigerant Recovery and Recycling Systems

Chapter 9

Refrigerant Recovery and Recycling Systems

Objectives

After completion of this chapter you should:

◆ *be more familiar with the safety procedures used when recovering/ recycling refrigerant.*

◆ *know more about refrigerant recovery/recycle units.*

◆ *know more about refrigerant recovery/recycle systems.*

◆ *know more about the disposal of used compressor oil.*

◆ *be more familiar with refrigerant recovery/recycle unit maintenance.*

◆ *be more familiar with the options concerning recovered refrigerants.*

◆ *know the different types of refrigerant recovery/recycle units available.*

In this chapter we discuss refrigerant recovery and recycling systems operation, maintenance, and safety. The type of system purchased will depend on personal requirements and desires. However, they all require periodic maintenance and care. The type of machine purchased determines the exact method that should be used to make the proper connections to the system being serviced. The manufacturer's recommendations should always be followed.

We will start with safety because it is most important to protect the equipment being serviced and human life. Be sure to follow all safety procedures to help in preventing personal injury or property damage.

Safety

Remember that the safe way is the only way. The equipment can be replaced but human life cannot. It should also be noted that only qualified refrigeration and air conditioning service personnel use this type of equipment. The following are some recommended steps to follow when operating refrigerant recovery, recycle, and reclaim equipment. Remember that this is not a complete safety list. These are only recommendations and do not take precedence over the equipment manufacturer's recommended procedures.

1. Use only DOT certified refillable cylinders.
2. Make certain that the cylinder is clean and has been evacuated.
3. Weigh both the refrigerant and the cylinder during the refilling process. Never overfill a cylinder. Use an accurate scale and fill the cylinder to only 80% of its rated weight at 70°F. If the cylinder is to be stored in a place that may reach more than 130°F, fill the cylinder to only 60% of its weight rating. Remember that service trucks get very hot when closed up during the summertime. If the cylinder is too full of liquid it could possibly explode because of the hydraulic pressure buildup during high temperatures. If a cylinder should explode there could be very serious injury, death, or property damage.
4. Never mix refrigerants. Make sure that any refrigerant in the cylinder is the same type that is being recovered from the system being serviced.
5. When refrigerant has been recovered into a cylinder, be sure to mark what type of refrigerant is in the cylinder. Use that cylinder only for the type of refrigerant that is marked on it.
6. When the recovered refrigerant is contaminated, it must be restored to ARI Standard 700-88 purity by either recycling or having it reclaimed. If this is not possible it must be properly destroyed.
7. Handle cylinders carefully. Do not drop or throw them off roof tops. They should be secured in the vertical position. Do not heat a cylinder with a welding torch. This weakens the cylinder wall allowing the cylinder to rupture much more easily.
8. When working with refrigerants the service technician should always wear goggles and rubber gloves with a thermal lining.
9. Make certain that there is plenty of ventilation in the equipment room when working with refrigerants because they will displace the oxygen in the air and possibly cause suffocation.
10. Always obtain and study the Material Safety Data Sheets (MSDS) for all chemicals with which you are working. These can be obtained from the refrigerant supplier.

131

11. Never pressurize a refrigeration system with oxygen. Oxygen and oil under certain conditions form an explosive mixture.
12. Make certain that the electrical disconnect switch is turned off before servicing any of the electrical components.
13. If moisture is allowed to enter the refrigeration system it will cause many problems and possible damage to the system. Keep all hoses, gauges, and other parts and tools used on the refrigeration system clean and dry.
14. Never operate a refrigerant recovery system in an explosive environment.

Refrigerant Recovery

Refrigerant recovery is one of the first things that is done to a refrigeration system that is not functioning properly. This recovered refrigerant may need to be cleaned or reclaimed before it is safe to use again. There are several different methods used to recover refrigerant, such as liquid pump recovery, use of equipment that is specially designed for refrigerant recovery. The equipment that is specially designed for recovery may still be further divided into those that recover vapor only, those that recover both liquid and vapor, those that separate the oil from the refrigerant, and those that do not separate the oil from the refrigerant. Each type has its advantages and disadvantages. The type chosen should be one that will serve the user's purpose.

The liquid pump is a much faster operation; however, the vapor left in the system must be removed with a vapor recovery unit. The vapor recovery unit removes all of the refrigerant, both liquid and vapor. It is a much slower process, but only one piece of recovery equipment is needed.

Regardless of which unit is chosen, all recovery methods are similar. The main thing is to have good tools and equipment. There must be no leaks in the gauge manifold, the gauge connections, the hoses, or their connections.

Recovery/Recycle Units

The recovery units that separate the oil from the refrigerant are not necessarily preferred to those without this function. It depends mainly on the desires of the purchaser. In many cases the oil taken from the system being serviced is contaminated and must be disposed of properly. If this contaminated oil is left in the refrigerant and put back into the system, then the new components are also somewhat contaminated. Remember that any oil taken from the system must be disposed of properly and in accordance with local, state, and federal regulations. The federal regulation is listed in Title 40 CFR 266 under subpart H-"Hazardous Waste Burned in Boilers and Industrial Furnaces" Section 266.107 "Standards to Control Hydrogen Chloride (HCl) and Chlorine Gas (Cl₂) Emissions." See Figure 9-1 for a schematic diagram of a system that separates oil from refrigerant during recovery.

Recovery/Recycle Systems

The recovery of refrigerant has been done for many years by responsible service technicians. This was first called refrigerant pump-out. One purpose of recovering refrigerants was to conserve large quantities of refrigerant rather than vent it to the atmosphere and install a complete new charge. It was done mainly to save the customer money. In most cases, not all of the refrigerant was recovered, but the major part of it was. Total refrigerant recovery could make the labor cost more than the refrigerant itself.

Two types of recovery/recycle equipment are available. They are the single pass and multiple pass units. Some recovery/recycle units combine both of these in one housing while others have them separated. These systems have different pumping capabilities. The average pumping time is about 2 pounds of refrigerant per minute.

The one chosen is determined by the majority of work or the location of the equipment. For example if the air conditioning or refrigeration system is located on the roof, the technician would not want to carry a heavy recovery unit up to the roof every time recovery service was needed. The desired refrigerant removal capabilities of the unit also must be consid-

133

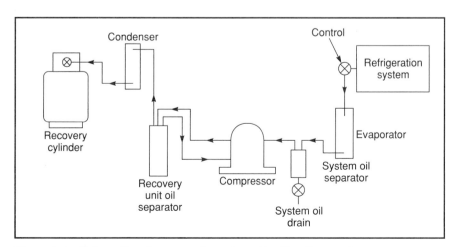

Figure 9-1   Schematic of recovery/recycle unit with oil separator.

ered. It would not be practical to purchase a unit that could remove 4 or 5 pounds of refrigerant per minute when most of the service provided is on systems that only hold 6 or 8 pounds of refrigerant.

Single Pass. The single pass recovery/recycling units pass the refrigerant through filter-driers and/or some type of distillation process. In these units the refrigerant makes only one pass through the internal system. The refrigerant is passed through the filter-driers, through the distillation process, and then into the recovery cylinder, Figure 9-2.

Multiple Pass. These types of units recirculate the refrigerant through the internal refrigeration circuit several times, or for a period of time. The refrigerant is circulated through filter-driers during this process. When the refrigerant is sufficiently cleaned, it is then pumped into the recovery cylinder, Figure 9-3.

Since recycling has become a factor in refrigerant conservation there have been many theories and comments made on whether recycled refrigerant is safe to use in systems. If we think back, service technicians have been cleaning systems with severe hermetic compressor motor burn-out using filter-driers with excellent results for many years. In most instances, when a repeat failure is experienced, the system just was not cleaned sufficiently. Often this is because the customer did not want to pay for sufficient time and materials to properly clean the system.

It should be remembered that when the oil is separated from the refrigerant most of the contaminants remain in the oil. Therefore, when the oil is removed from the system the refrigerant is much easier to clean. Most of the recovery/recycle units use some type of filter-drier to remove moisture, acids, and particulate matter from the refrigerant. When

134

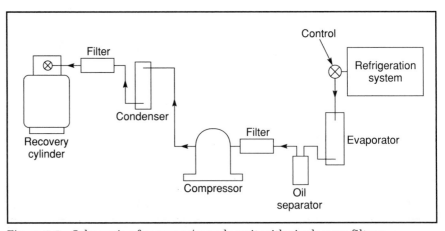

Figure 9-2   Schematic of recovery/recycle unit with single pass filters.

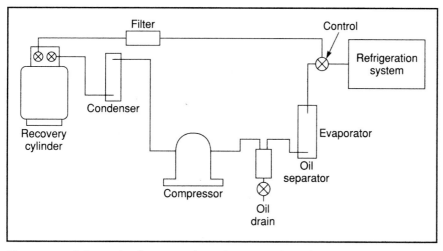

Figure 9-3   Schematic of recovery/recycle unit with multiple pass filters.

refrigerant has received this type of treatment, it is sufficiently clean to place back into the system.

Note: Recovery/recycle units do not remove noncondensables or separate different types of refrigerants. When these contaminants are in the system other means must be used to make the refrigerant suitable for reuse. When noncondensables are present they can usually be purged off after the recovery cylinder has cooled and the noncondensables have risen to the top of the cylinder. When different types of refrigerants are mixed it is usually about as economical to have them properly disposed of as it is to reclaim them.

135

Recovery/Recycle Unit Use

It is always best to follow the unit manufacturer's recommendations when using any type of equipment. The manufacturer designed and made the equipment, and should know how to get the best results from it.

The following are some suggestions for using recovery/recycle equipment. Be sure to follow the equipment manufacturer's recommendations.

1. Most recovery/recycle units need to be evacuated before using.
2. Most of these units need to be evacuated when recovering a different type of refrigerant.
3. Be sure to use either an empty recovery cylinder or one that contains the same type of refrigerant being recovered.

4. If the recovery cylinder is empty, the process will proceed much more rapidly if it is evacuated before the recovery process begins.
5. Be sure to purge the hoses and all connections of air.
6. Be sure that new filter-driers are installed on the recovery system.
7. If the recovery system has an oil separation function, be sure that the oil separator is empty.
8. When the recovery process is completed, drain the oil from the oil separator into a graduated container so that the exact amount can be replaced into the system. Also, this provides an easier way of collecting the oil so that it can be disposed of properly. Use caution when handling used refrigeration oil, especially oil that has come from a system with a burned hermetic compressor motor. This oil can be very toxic and acidic and can cause acid burns.
9. Test the refrigerant with an acid test kit, or a Totaltest unit to determine its acid and moisture content. Do not place questionable refrigerant into the system. Charge the system with new or reclaimed refrigerant and have the original refrigerant reclaimed.
10. To speed up the recovery process, use hoses and connections with the largest inside diameter possible to reduce the restriction offered by them.
11. When Schrader valves are installed on the system, remove their cores to reduce the restriction at this point.
12. Use the shortest hoses possible.

Recovery/Recycle Unit Maintenance

The maintenance procedures recommended by the unit manufacturer should be completed at the prescribed intervals. The following are some suggestions for maintenance.

Compressor Oil

Drain the compressor oil after recovering the refrigerant from every system with a burned out hermetic compressor motor. To prevent mixing the refrigerants the oil should be changed before the recovery of a different type of refrigerant. During normal operation, oil is carried out of the compressor at a rate of about $1/2\%$ by weight. Check the oil before every use and maintain the correct level to prevent damage to the compressor.

Filter-Driers

To prevent mixing of refrigerants, change the filter-drier before recovering a different refrigerant. Also, change the filter-drier after recovering a contaminated refrigerant. Evacuate the recovery unit and the connecting hoses before recovering more refrigerant.

Recovered Refrigerant Options

Basically there are four options available when working with recovered refrigerants. They are recover and reuse, recover and recycle on-site, recover and reclaim off-site, and recover and destroy.

Recover and Reuse

This procedure is normally done on-site. The recovered refrigerant has been tested and found to be in reusable condition. After the desired repairs or maintenance has been completed, the same refrigerant can be recharged into the system without any further cleaning. Care must be taken to prevent contamination during the recovery process.

Several things must be considered when this procedure is used. The contractor assumes: (1) full liability for the condition of the refrigerant, (2) possible equipment warranty problems, (3) that all regulations have been followed, and (4) that no refrigerant has been lost.

137

Recover and Recycle On-Site

When the recovered refrigerant is found to be only mildly contaminated, the refrigerant may be cleaned on-site to sufficiently meet the cleanliness required. This method is best suited where the refrigerant charge is small and the refrigerant can be cleaned in a relatively short period of time.

Among the considerations for on-site recycling are: the contractor assumes full liability for the condition of the refrigerant; the contractor makes certain that all regulations have been followed; since there is no laboratory analysis of the refrigerant, there can be no certification; and the possibility of equipment warranty problems.

Recover and Reclaim Off-Site

Sometimes the refrigerant is so contaminated that only reclaimed or new refrigerant is the safest method to use. In some instances the owners of large systems require that the purity of the refrigerant charged into the system be certified by a reclaim facility. Be sure that when reclaimed refrigerant is used, certification is obtained for that refrigerant. This relieves the contractor of any responsibility for the condition of the refrigerant. It is a good idea to know personally the qualifications and respectability of the person operating the reclaim facility.

An off-site recovery facility assumes full responsibility for the condition of the refrigerant. Use of such facilities means the refrigerant is readily available for use, and the cost of new refrigerant has been avoided.

Recover and Destroy

Sometimes the refrigerant is so contaminated that it cannot be reclaimed. This usually occurs when refrigerants have been mixed. Most other types of contaminants can be removed during a proper reclaim procedure.

The facility used to destroy contaminated refrigerants must meet strict federal regulations. CFCs are destroyed only by incineration, which requires that the facility be able to capture the fluorine particles that are produced during the incineration process. This is a very expensive option, but sometimes it is necessary.

138

Important considerations when the destroy process is chosen are: new and expensive refrigerant must be purchased, the disposal of the refrigerant is expensive, and the refrigerant is lost forever.

Recovery, Recycling, and/or Reclaiming Equipment Standards

ARI Standard 740 is the standard to which these units must operate. This list is constantly being updated. Contact ARI for the most recent update.

Summary

◆ The type of recovery/recycle unit purchased depends on personal requirements and desires, but all require periodic maintenance and care.

◆ The safe way is the only way. The equipment can be replaced but human life cannot. Only qualified refrigeration and air conditioning personnel use this type of equipment.

◆ Refrigerant recovery is one of the first things done to a refrigeration system that is malfunctioning.

◆ Recovered refrigerant may need to be cleaned and recleaned before it is safe to use again.

◆ Methods used to recover refrigerant include liquid pump recovery, making use of equipment that is specially designed for liquid refrigerant recovery.

◆ The equipment specially designed for recovery may be divided further into those that recover vapor only, those that recover both liquid and vapor, those that separate the oil from the refrigerant, and those that do not separate the oil from the refrigerant.

◆ Regardless of the unit chosen, all recovery methods are similar.

◆ Oil taken from the system being serviced often is contaminated and must be disposed of properly.

◆ The two main types of recovery/recycle equipment are single pass and multiple pass units.

◆ Single pass recovery/recycle units pass the refrigerant through filter-driers and/or some type of distillation process.

◆ Multiple pass recovery/recycle units recirculate the refrigerant through the internal refrigeration circuit several times, or for a specified length of time. The refrigerant is circulated through filter-driers during this process. When the refrigerant is sufficiently cleaned, it is then pumped into the recovery cylinder.

◆ When the oil is separated from the refrigerant, most of the contaminants remain in the oil.

◆ Most recovery/recycle units use some type of filter-drier to remove moisture, acids, and particulate matter from the refrigerant. When the refrigerant has received this type of treatment, it is sufficiently clean to place back into the system provided there are no noncondensables or other types of refrigerant present.

◆ Drain the recovery/recycle unit compressor oil after recovering the refrigerant from every system with a burned-out hermetic compressor motor.

◆ To prevent mixing refrigerants, the recovery/recycle unit compressor oil should be changed before the recovery of different types of refrigerant. Check the oil before every use and maintain the correct level to prevent damage to the compressor.

◆ To prevent mixing refrigerants, change the filter-driers before recovering a different type of refrigerant. Change the filter-driers after recovering contaminated refrigerant.

◆ Evacuate the recovery unit and the connecting hoses before recovering more refrigerant.

139

◆ Four options are available when working with recovered refrigerants: recover and reuse, recover and recycle on-site, recover and reclaim off-site, and recover and destroy.

◆ Key Terms ◆

filter-drier

multiple pass

recovery/recycle systems

recovery/recycle units

refrigerant recovery

single pass

Review Questions

Essay

1. What do all recovery/recycle units require?

2. What is the only way to work?

3. What is the main thing to have when recovering refrigerant?

4. What must be done with contaminated compressor lubricating oil?

5. Name the two types of recovery/recycle equipment available.

Fill-in-the Blank

6. Only _____ refrigeration and air conditioning personnel should use recovery/recycle equipment.

7. Regardless of the unit chosen, all recovery _____ are similar.

8. The _____ _____ recovery/recycle unit passes the refrigerant through filter-driers and/or some type of distillation process.

9. Filter-driers are used to remove _____, _____, and _____ _____ from the refrigerant.

10. The filter-driers should be changed before recovering a different type of refrigerant to prevent _____ the refrigerants.

True-False

11. The type of recovery/recycle unit purchased depends on the desires of the customer.

12. Recovered refrigerant may need to be cleaned and recleaned before it is safe to use again.

13. Multiple pass recovery/recycle units recirculate the refrigerant through the internal refrigeration circuit several times.

14. Most contaminants in a system remain in the oil.

15. Filter-driers remove noncondensables and moisture.

Multiple Choice

16. The procedure used after recovering refrigerant from a system with a burned-out hermetic compressor motor is
    a. save the refrigerant for future use.
    b. drain the recovery/recycle unit compressor oil.
    c. flush the system with CFC-11.
    d. pressurize the system with HCFC-22.

17. One option with recovered refrigerant is to
    a. recover and recycle on-site.
    b. store in a recovery cylinder with other refrigerants.
    c. test on-site for purity.
    d. make sure that all the oil is removed from it.

18. The contaminated oil taken from a system
    a. is poured down the drain.
    b. is put back in the system.
    c. is properly disposed of.
    d. is left by the unit.

141

19. The two basic types of recovery/recycle units available are
    a. single and multiple pass.
    b. single pass and reclaim.
    c. multiple pass and reclaim.
    d. recover and reclaim.

20. The following *is not* an option when working with recovered refrigerants:
    a. recover and recycle on-site.
    b. recover and reclaim off-site.
    c. recover and destroy.
    d. check for purity on-site.

APPENDIX A
Answers to Review Questions

PART I

Chapter 1

1. chlorofluorocarbons (CFCs) and halons
2. the stratospheric ozone layer
3. stratospheric ozone and atmospheric ozone
4. Atmospheric ozone
5. T
6. F
7. T
8. F
9. CFC-11, CFC-12, and CFC-114.
10. 1.5 to 2
11. 20
12. fish
13. C
14. C
15. A

Chapter 2

1. the evaporation of a liquid
2. ammonia
3. chlorofluorocarbons (CFCs)
4. chlorofluorocarbons (CFCs)
5. chlorofluorocarbons (CFCs)
6. They have no hydrogen in their composition.
7. They have a very high ozone depletion potential (ODP).
8. hydrofluorocarbon (HFC) and hydrochlorofluorocarbon (HCFC) refrigerants
9. hydrofluorocarbon
10. hydrogen
11. hydrogen

12. HFCs and HCFCs; smog

13. oil

14. HFC-134a; CFC-12

15. ternary; CFC-12

16. F

17. T

18. F

19. F

20. F

21. T

22. A

23. A

24. C

25. B

26. C

27. A

28. C

29. A

30. D

143

Chapter 3

1. in 1978

2. aerosol propellant applications

3. September, 1987

4. January, 1989

5. New measurements of the loss of the ozone layer were much greater than the computer models had predicted.

6. the Clean Air Act of 1990

7. T

8. F

9. F

10. F

11. I

12. 1986

13. stringent

14. complete

15. taxes

16. A
17. C
18. D
19. B
20. B
21. A
22. D
23. C
24. D

Chapter 4

1. by 1995
2. $3.10 per pound
3. the technician
4. one kilogram (2.2 pounds)
5. $25,000 per day per offense
6. keeping the refrigerant in the system
7. with the recovery unit
8. with a recovery unit
9. It should be dried.
10. It should be recycled.
11. F
12. F
13. T
14. T
15. T
16. refrigerant management system
17. purge units
18. frequently
19. purge unit
20. pressure relief

Chapter 5

1. It should be cleaned (recycled).
2. It must be reclaimed to ARI 700-8 specifications.
3. a Totaltest refrigerant tester
4. recycling

5. It must be reclaimed to ARI 700-88 specifications.
6. noncondensables
7. reclaiming
8. recovered
9. recycled
10. recycling
11. T
12. T
13. F
14. F
15. T
16. C
17. D
18. A
19. B
20. A

145

Chapter 6

1. the specific data sheet for the product
2. epinephrine
3. because CFCs displace the air from enclosed or semienclosed spaces
4. because they will decompose forming hydrochloric and hydrofluoric acid vapors
5. by weight
6. The refrigerant should be identified.
7. It should be marked and labeled for the type of refrigerant.
8. Never mix refrigerants in the same container.
9. to verify that the test date has not expired
10. to 10% of the drum height
11. 80% at 70°F
12. 300 psig
13. Empty tag
14. serial number
15. five
16. damaged
17. gross weight
18. monitor

19. leaking
20. cover
21. F
22. F
23. T
24. T
25. F
26. F
27. F
28. T
29. F
30. F
31. A
32. C
33. B
34. B
35. C

146

Part II

Chapter 7

1. HFC-123
2. from 2% to 5%
3. a full equipment room monitoring system
4. those that originally contained CFC-11 or CFC-113 (excluding those originally used for cleaning agents)
5. yes
6. yes
7. gray body and a yellow top
8. at least a 29-inch vacuum Hg
9. either HCFC-22 or HFC-134a
10. no
11. HCFC-22
13. HFC-134a
14. lubricating oil
15. HFC-134a
16. T
17. T

18. T
19. F
20. F
21. A
22. C
23. B
24. D
25. B

Chapter 8

1. low-pressure systems, high-pressure systems containing more than 50 pounds of refrigerant, high-pressure systems containing less than 50 pounds of refrigerant, and small appliances
2. a service aperture
3. the line could rupture and the refrigerant could be overheated producing hydrochloric acid
4. It must pump a vacuum of approximately 10 inches to 20 inches Hg.
5. $25,000 per day per offense
6. safe disposal
7. overheat
8. cylinder, bag
9. 60%
10. air, oxygen
11. T
12. F
13. F
14. F
15. T
16. B
17. D
18. A
19. A
20. C

147

Part III

Chapter 9

1. periodic maintenance
2. the safe way

3. good tools and equipment
4. properly disposed of
5. single pass and multiple pass
6. qualified
7. methods
8. single pass
9. moisture, acid, particulate matter
10. mixing
11. F
12. T
13. T
14. T
15. F
16. B
17. A
18. C
19. A
20. D

148

APPENDIX B
Glossary

A

ARI (Air Conditioning and Refrigeration Institute): An association for refrigeration and air conditioning personnel. ARI does research and produces educational materials and standards for the industry.

ASME (American Society of Mechanical Engineers): An organization to which all engineers can belong. Through ASME, engineers can keep current on new procedures in the engineering field.

Atmospheric ozone: As the ultraviolet rays from the sun reach the earth they combine with smog and pollution and cause atmospheric ozone. This ozone is harmful, and it must not be confused with the stratospheric ozone layer.

C

Cardiac arrhythmia: An irregularity in the rhythm of the heart's beating. The heart may speed up quickly then slow down for no apparent reason.

CFC (chlorofluorocarbon): Refrigerants that contain the atom chlorine, which attacks the stratospheric ozone layer.

CFR (code of federal regulations): Regulations that are generated and published by a branch of the U.S. government. They must be adhered to for any undertaking regulated by the Federal Government.

Class I substances: These are fully halogenated chlorofluorocarbon refrigerants and are controlled by the Clean Air Act.

Class II substances: This group consists of the halons that are used as fire extinguisher agents. They are controlled by the Clean Air Act.

D

DOT (U. S. Department of Transportation): A federal agency charged with the regulation and control of all hazardous materials shipped through commercial means.

DOT 39 (disposable refrigerant cylinder): These cylinders should never be refilled or used as compressed air tanks.

E

EPA (Environmental Protection Agency): An agency of the U.S. government charged with the responsibility of protecting the environment and enforcing the Clean Air Act of 1990.

F

filter-drier: A device that is placed in a refrigerant line to help remove moisture, acid, and particulate matter from the refrigerant.

FDA (U.S. Food and Drug Administration): An agency of the U.S. government charged with the responsibility to govern the foods and drugs available to the U.S. citizens.

G

greenhouse effect: The retention of heat from sunlight at the earth's surface caused by atmospheric carbon dioxide that admits shortwave radiation emitted by the earth.

H

HCFC (hydrochlorofluorocarbon refrigerant): Refrigerants that contain the chlorine atom and the hydrogen atom which causes the chlorine atom to dissipate more rapidly in the atmosphere.

HFC (hydrofluorocarbon): A chemical refrigerant that does not contain the chlorine atom. It has an ODP of 0.

150

L

LAL (lowest achievable level): The lowest amount of emissions possible of CFC and HCFC refrigerants.

M

Montreal Protocol: An agreement signed by the participating parties who have agreed to reduce CFC and HCFC emissions into the atmosphere.

multiple pass: This term applies to a recovery/recycle unit that removes the refrigerant from the system being repaired and circulates through the recovery/recycle system several times or for a time to remove contaminants other than noncondensables and other types of refrigerants. The refrigerant is then pumped into the recovery cylinder.

N

National Institute for Standards and Technology: A national institute that helps set standards and technology for mobile air conditioning systems.

O

ODP (ozone depletion potential): See ozone depletion weight.

ODW (ozone depletion weight): A number assigned to a chemical representing the ability of the substance to destroy the stratospheric ozone layer. The ozone depletion weight is the same as the ODP (ozone depletion potential).

ozone (O_3): An unstable pale-blue gas, with a penetrating odor; it is an allotropic form of oxygen formed usually by a silent electrical discharge in the air.

ozone layer: This is a layer of ozone that is in the atmosphere from just a few to several miles above the earth. It is what protects the earth from the ultraviolet solar radiation rays from the sun.

P

PAFTT (Program for Alternative Fluorocarbon Toxicity Testing): A testing agency formed by several refrigerant manufacturers to study and make toxicity tests on new refrigerants.

PAG (polyalkylene glycol): A synthetic lubricating oil generally used with HFC-134a refrigerant.

R

reclaim: To reclaim a refrigerant is to reprocess it to new specifications by means that may include distillation. Reclaim requires a chemical analysis of the refrigerant to determine that appropriate product specifications are met. This term usually implies the use of processes or procedures available only at a reprocessing facility.

151

recovery: The recovery of a refrigerant is to remove refrigerant in any condition from a system and store it in an external container without necessarily testing or processing it in any way.

recovery/recycle systems: The refrigerant circuit inside the recovery/recycle unit. It is used to transfer the refrigerant from the system being repaired to the recovery cylinder.

recovery/recycle unit: This is the complete unit used to recover/recycle the refrigerant from the system being repaired.

recycle: To recycle means to clean the refrigerant for reuse by oil separation and single or multiple passes through devices, such as replaceable core filter-driers, which reduce moisture, acidity, and particulate matter. This term usually applies to procedures implemented at the field job site or at a local service shop.

RMS (refrigerant management system): A device that recovers the refrigerant charge (including the vapor) of negative-pressure chillers, recycles it, then returns it to the chiller.

S

single pass: This term applies to a recovery/recycle system that removes the refrigerant from the system being repaired and passes it through a system of filter-driers, and perhaps a distillation process, and then into the recovery cylinder.

SSU (Saybolt Seconds Universal): The method used in determining the flow rate of fluids. It is usually used to designate the weight of lubricating oil.

stratospheric ozone layer: The atmospheric zone above the earth, extending from six to fifteen miles above the earth's surface. It protects the earth from ultraviolet rays from the sun.

T

TLV (threshold limit value): The concentration of refrigerant in air above which a person can become drowsy or have a loss of concentration.

Totaltest tester: This is a device used to easily check for moisture and acid in a refrigerant.

U

ultraviolet radiation: The invisible rays from the sun that have damaging effects on objects on the earth. This is what can cause sunburn when a person sunbathes.

UL (Underwriters Laboratories): An independent laboratory that operates and performs testing on devices, systems, and materials.

V

VOC (volatile organic compound): Chemicals that are negligibly photochemically reactive.

INDEX

153

154

155